TURNING POINTS

IN

Military History

D1211138

TURNING POINTS
IN
Military History

William R. Weir

CITADEL PRESS
Kensington Publishing Corp.
www.kensingtonbooks.com

CITADEL PRESS BOOKS are published by

Kensington Publishing Corp.
850 Third Avenue
New York, NY 10022

First printing: March 2005

10 9 8 7 6 5 4 3 2 1

Printed in the United States of America

Library of Congress Control Number: 2004113768

ISBN 0-8065-2627-0

Contents

Introduction

WHEN WE THINK of the turning points in war, the first thing that occurs to most people is the introduction of new weapons. The use of airplanes certainly caused a turning point, although the result was quite different from what either the apostles of air power or their detractors foresaw. The apostles—Giulio Douhet, Billy Mitchell, Hugh Trenchard and like-minded generals—were sure that bombers would render traditional armies and navies practically superfluous. Their detractors believed in their hearts that if God had wanted people to fly, he would have given them wings. In this case, the detractors got their way. The result was "strategic bombing," which killed a lot of civilians, but neither broke the will of any enemy population nor did much, in the case of Germany, at least, to diminish enemy war production. And it certainly did not render armies and navies unnecessary.

More often, the detractors win. In 1916, Britain's Sir Douglas Haig, who thought two machine guns per battalion were more than sufficient, had his cavalry ready to exploit the breakthrough on the Somme that never came while he fed his men into the German machine guns. Sir Douglas was a hard man to convince. He didn't stop his offensive until he had lost more men in that battle than the United States lost in all of World War II. Unfortunately, Haig wasn't alone. He was typical of World War I generals. The resultant slaughter was a turning point not only in warfare but in world history.

Europeans do not have a patent on this sort of blindness. In the American Civil War, generals on both sides used the smoothbore tactics they had learned in their youth. The trouble was that nobody used smoothbores anymore. Rifles had ten times the smoothbores' effective range. But the generals sent their troops marching shoulder-to-shoulder against entrenched enemies who could hit them from a half-mile away. So more Americans were killed in the Civil War than in any other war we've fought. And the U.S. population in 1860 was barely more than a tenth of what it is today.

Not all turning points are created by new weapons. Sometimes it's a matter of attitude. And attitudes have many strange roots. For example, some historians blame the pugnaciousness of average Europeans, compared with average Chinese or Indians, on the climate. Europeans kept a lot of livestock, but the European winter did not provide enough fodder for their herds and flocks. Consequently, every fall, they slaughtered most of their pigs, cows and sheep and pickled or dried the meat. So, according to the theory, they were not averse to bleeding human enemies, too. And actually, European merchants were not as passive as those of the Orient. They walled their cities and fought the nobles. Both they and the nobles also hired mercenaries, most of them runaway serfs. All that fighting led to an all-European arms race, which led to Europeans developing weapons that let them dominate the world.

Even earlier, the pugnacious independence of minor European nobles led to the construction of tens of thousands of castles, an abundance of forts seen nowhere else in the world. It seems that abundance, more than anything else, was what discouraged the Mongols of Genghis Khan and his successors from extending the Great Khan's empire to the Atlantic Ocean. That abundance also led European monarchs to invest heavily in cannons—the best way to teach upstart barons that castles cannot protect defiance.

Cannons in this case greatly strengthened the notion that kings rule by divine right. So did the age of limited warfare. The kings, seeing that the kind of total war waged in the Thirty Years' War impoverished every nation that participated, created standing armies. They created them by "recruiting" (usually by press gangs) the least productive members of society—the drunks, vagabonds and petty criminals. They trained these men until they became virtual automatons.

Then the kings played with their armies the way small boys play with toy soldiers. This cold-bloodedness merely reflected the attitude of rulers toward everything. The kings had a monopoly on the use of force, and they used it any way they liked. In the end, that provoked a reaction—revolution. The French Revolution swept away the regular armies of the ancien regime and introduced the notion of the nation at war. The notion is still with us.

War is about as undesirable as any large-scale human activity can be. But it's been with us since before history, and it will no doubt be with us for a long time to come. That's why it's important to understand it—not only weapons and tactics, but all the factors that lead to the turning points in warfare. For instance, we can learn now, in Iraq, what we should have learned in Vietnam—that the smartest weapons do not guarantee victory. If the enemy population does not accept defeat, you can't win until you kill every last person in that enemy population. No war is likely to reach that point, but the sort of slaughter that even approaches it is repugnant to all civilized people and likely to cause the victor far more trouble in the long run. Nationalism, which grew in Europe during the Napoleonic Wars, has spread to the so-called Third World and become even more intense. It is a factor—like climate, terrain, and even weaponry—that we must consider. We have reached a new turning point in military history that we have just begun to grapple with.

Timeline

The Basics

Oldest surviving complete spear dates from 250,000 B.C.

Bow and arrow used in North Africa about 15,000 B.C.

Earliest known swords found in Anatolia, dating from the late third millennium B.C.

Urbanite Warfare

Creation of Greek phalanx   **700 B.C.**

Battle of Marathon, Greek phalanx defeats Persians   **490 B.C.**

Gauls under Brennus sack Rome   **390 B.C.**

Battle of Leuctra, Thebans defeat Spartan phalanx   **371 B.C.**

Battle of Chaeronaea, Philip II conquers Greece   **338 B.C.**

Battle of Arbela (Gaugamela), Alexander the Great conquers Persia   **331 B.C.**

Battle of Bovianum, Roman legion decisively defeats Samnites   **305 B.C.**

Romans decisively defeat Macedonians at Pydna   **168 B.C.**

Goths destroy Roman army and kill Emperor Valens at Adrianople   **A.D. 378**

Gothic king Alaric sacks Rome   **A.D. 410**

The Rulers of the Plains

Sumerians invent two-wheeled chariot   **3000 B.C.**

Chariot-driving Aryans invade Iran   **2000 B.C.**

Aryans invade India   **1800 B.C.**

Hyksos complete conquest of Lower Egypt   **1700 B.C.**

Egyptians adopt Hyksos chariot and drive out Hyksos   **1580 B.C.**

Nomads begin riding horses   **1300 B.C.**

Kingdom of Parthia founded   **247 B.C.**

Parthian cavalry wipes out Romans under Crassus at Carrrhae   **53 B.C.**

Huns defeat Goths and drive them into Roman Empire   **A.D. 376**

Seljuk Turks overrun southwest Asia, including Anatolia   **1043–1071**

Battle of Hattin, crusader army annihilated by Turkish and Arab horse
   archers   **1187**

First Mongol invasion of Europe   **1223**

Second Mongol invasion of Europe   **1237**

Stonewalling the Nomads

Muslims from Spain make first raid into France   **712**

Charles Martel defeats Muslims near Tours   **732**

Vikings make first raid on Britain   **789**

Vikings begin raiding Frankland (France and Germany)   **809**

Magyars make first raid into western Europe   **862**

Doue-la-Fontaine, first stone castle, built   **950**

Stone castles begin replacing wooden strongholds in France, then all
   Europe   **950**

Battle of Clontarf, defeat of the last major Viking raid   **1014**

William the Conqueror brings portable wooden castle in his invasion of
   England   **1066**

Normans build wooden, then stone castles all over England from   **1066**

First Mongol invasion of Europe   **1232**

Second Mongol invasion of Europe   **1237**

Private Armies and Public Policy

Rise of Italian civic militias   **ca. 1100**

Catalan company defeats French cavalry in Sicily   **1282**

Black Death begins, resulting in labor shortage and freedom for
   serfs   **1346**

Mercenary companies replace civic militias in Italian cities from   **1350**

Venice, beginning inland campaign, hires condottiere with long-term
   contracts   **1405**

Beginning of Hundred Years' War, a breeding ground for
   mercenaries   **1450**

Francesco Sforza, a condottiere, takes over Milan   **1450**

The Devil's Snuff: The Gunpowder Revolution

Gunpowder in use in China   **1000**

Chinese use crude hand grenades   **1100**

Roger Bacon describes gunpowder   **1252**

First Mongol attempt to conquer Japan   **1274**

Second Mongol attempt to conquer Japan   **1281**

Chinese use bamboo guns   **1290**

Handguns in use at Florence   **1324**

Walter de Milemete manuscript shows picture of gun   **1326**

Chinese picture of gun   **1332**

Guns used at Crecy   **1346**

Fall of Constantinople   **1453**

Artillery ends Hundred Years War   **1453**

Siege of Pisa   **1500**

Battle of Bicocca   **1524**

Handgunners defeat French at Pavia   **1525**

Command of the Sea: War Under Sail

Caesar sees Gaulish ships sail into wind   **55 B.C.**

Greek fire used by Byzantine galleys   **A.D. 660**

Rudders begin to replace steering oars   **1050**

Cogs, north European merchant ships grow big enough to fight
   Viking ships   **1100**

Modern compass in use   **1269**

Nautical charts in use   **1270**

Cogs enter the Mediterranean   **1304**

Chinese "Treasure Fleet" of Zheng explores East Indies, India,
   Arabia and Africa   **1405–1433**

Portuguese caravels reach mouth of the Gambia in Africa   **1450**

Bartomeu Dias rounds Cape of Good Hope   **1487**

Christopher Columbus reaches America   **1492**

Battle of Diu ends Muslim control of Indian Ocean   **1509**

The Bloody Birth of Standing Armies

Protestant Reformation begins   **1517**

Lutheran peasant revolt crushed   **1524**

Long series of Huguenot Wars begins in France   **1562**

Dutch-Spanish War (80 Years' War) begins   **1585**

Maurice of Nassau's new model army opens offensive   **1590**

Thirty Years' War begins   **1618**

Ernst von Mansfeld organizes army that lives off the land. Others
follow his example   **1621**

Wallenstein's army destroys Mansfeld's   **1626**

Wallenstein defeats Danes and knocks them out of war   **1629**

Sweden enters war   **1630**

Gustavus defeats Tilly at Breitfeld   **1631**

Tilly killed battling Gustavus's army at the Lech   **1632**

Gustavus defeats Wallenstein at Lützen and is killed   **1632**

Wallenstein assassinated   **1634**

Swedish army heavily defeated by Imperialists   **1634**

Imperialists heavily defeated by Swedish army   **1642**

Peace of Westphalia, end of Thirty Years' War   **1648**

Quiet Revolution: The Age of Limited Warfare

The Viscount of Turenne, a master of maneuver defeats Allies
at Sinsheim   **1674**

Turenne again defeats Allies, but is killed in battle soon after   **1675**

War of the League of Augsburg begins   **1688**

Socket bayonet invented   **1688**

Francois of Luxembourg heavily defeats allies at Fleurus   **1690**

War of the League of Augsburg ends with return to status quo   **1697**

War of Spanish Succession (Queen Anne's War in America)
begins   **1701**

Duke of Marlborough, an untypical general of the age because he
actively sought battle, got more battle than he expected at
Malplaquet—he won, but his army's casualties were so great he was
removed from command   **1709**

War of Spanish Succession ends, with Louis XIV's grandson still king
of Spain   **1714**

War of the Polish Succession begins   **1733**

War of the Polish Succession ends   **1734**

War of Jenkins' Ear, which became War of the Austrian Succession, begins **1739**

War of the Austrian Succession—Prussia, Bavaria, France, Saxony, Savoy and Sweden vs. Austria, England and the Netherlands— begins **1740**

Battle of Chotusitz, Frederick the Great defeats Austria, suffers 7,000 casualties to Austrians' 3,000. Austria gives Prussia Silesia **1742**

Battle of Fontenoy **1745**

Battle of Culloden: English defeat Scottish rebels, institute reign of terror **1748**

End of War of Austrian Succession. Prussia re-entered war, which ultimately involved Spain, Sardinia, and Genoa in addition to original participants and became King George's War as well as War of Jenkins' Ear in America **1748**

Beginning of French and Indian War, which became Seven Years' War in Europe and India. War in Europe involved soldiers alone; war in America marked by massacres of civilians, torture of prisoners and cannibalism. **1754**

End of Seven Years' War with France practically driven out of North America **1763**

Boston Massacre **1770**

Boston Tea Party **1773**

Battles of Lexington and Concord **1775**

Declaration of Independence **1776**

Battle of Saratoga; Burgoyne surrenders, France declares war on Britain **1777**

British evacuate Philadelphia and occupy New York; Battle of Monmouth **1778**

French fleet and American troops attack Newport **1778**

Walter Butler's Tories and Joseph Brant's Mohawks massacre settlers in Cherry Valley. Pennsylvania Quakers fail to warn Connecticut settlers in the valley. American troops massacre Indian villages in Pennsylvania and New York. **1778**

Spain declares war on Britain **1779**

British capture Charleston **1780**

British under Cornwallis defeat Gates at Camden **1780**

American militia rout Tories under Patrick Ferguson at King's Mountain **1780**

Americans under Daniel Morgan annihilate Bannestre Tarleton's legion at Cowpens **1781**

Charles Cornwallis's British defeat Americans at Guilford Courthouse, but, unable to resupply his army, he moved to Yorktown, Virginia. American and French troops besiege him while French fleet cuts off help from the sea. He surrenders. **1781**

American Revolution, actually a world war involving the United States, the United Kingdom, France, Spain and the Netherlands, ends **1783**

Fall of the Bastille, beginning of French Revolution and a new turning point in military history **1789**

Not Armies, but Nations

Declaration of the Rights of Man published **1791**

Slave rebellion in Haiti begins **1791**

French monarchy abolished **1792**

Allies invade France, but retreat after non-battle at Valmy **1792**

France introduces conscription. French armies defeat British, Dutch, and Austrians **1793**

Colonel Napoléon Bonaparte promoted to brigadier general **1793**

Anti-French allies again invade France and are again defeated **1794**

French forces occupy the Netherlands **1795**

Prussia, Spain, Saxony, Hanover and Hesse-Cassel leave anti-French alliance **1795**

Napoléon Bonaparte invades Italy and defeats Austrians **1796**

Austria accepts Napoléon's peace terms **1797**

French set up Roman Republic in Rome and Helvetian Republic in Switzerland **1798**

French win Battle of the Pyramids; British, Battle of the Nile **1798**

Napoléon Bonaparte becomes first consul of France **1799**

British adopt shrapnel shell **1803**

French forces leave Haiti **1803**

Napoléon becomes emperor of France **1804**

Battle of Austerlitz: Napoléon destroys vast allied army **1805**

French armies take Berlin and Warsaw **1806**

French take Danzig and defeat Russians at Friedland **1807**

British fleet burns Copenhagen with shells and rockets **1807**

French install Joseph Bonaparte as king of Spain. Guerrilla war begins. British under Arthur Wellesley (later Duke of Wellington) land to aid Spaniards. Napoléon takes personal command. British driven out of Spain. **1808**

British and French fight indecisively in Portugal while Napoléon repeatedly defeats Austria and forces Austria to surrender   **1809**

Napoléon defeats Russians at Borodino, retreating Russians burn Moscow, French begin long disastrous retreat   **1812**

War of 1812 begins   **1812**

Americans take York (modern Toronto) and burn public buildings   **1813**

Napoléon raises new army and attacks allies in Germany; after several victories, he is defeated heavily at Leipzig   **1813**

British and Spanish drive French out of Spain   **1813**

Americans defeat British on Lake Erie, but are defeated in attempt to take Montreal; British and Indians take Fort Niagara and burn Buffalo   **1813**

Allies invade France; Napoléon abdicates and is exiled; British capture Washington, burn White House and Capitol   **1814**

British reinforce army in North America and move south from Montreal; their naval forces are defeated on Lake Champlain and they return to Montreal   **1814**

Napoléon returns from exile; Battle of Waterloo; Napoléon exiled again; old regimes reinstated   **1815**

British repulsed at Baltimore try new landing at New Orleans, resulting in most lopsided British defeat in modern history   **1815**

The American Civil War

Confederates take Fort Sumter; defeat Union at first Battle of Bull Run   **1861**

First battle between ironclads; U. S. Grant's Union forces defeat Confederates at Shiloh; Farragut's fleet takes New Orleans, Baton Rouge and Nashville; McClellan's Union army forced back at Seven Days Battle; Union army defeated at Second Battle of Bull Run; Robert E. Lee's first invasion of the North defeated by George McClellan at Antietam, the bloodiest single day of the war   **1862**

Abraham Lincoln signs Emancipation Proclamation; Lee's second invasion of the North defeated at Gettysburg; Grant takes Vicksburg; Confederates' failure to win a crushing victory at Chickamauga broke their morale, according to Gen. D. H. Hill, resulting in the rout of Chattanooga   **1863**

Grant's armies begin closing in on Richmond; Sherman takes Atlanta and marches to the sea; George Thomas's troops wipe out John B. Hood's Army of Tennessee   **1864**

Lee surrenders; Lincoln assassinated; war ends   **1865**

War by the Timetable

Schleswig-Holstein War: Prussia and Austria vs. Denmark **1864**

Seven Weeks' War: Prussia vs. Austria **1866**

Franco Prussian War begins **1870**

Napoleon III surrenders **1870**

Besieged Paris surrenders **1871**

Out of Africa

First Boer War begins; British defeated at Bronkhorst Spruit **1880**

Further British defeats at Laing's Nek, Schuin's Hoogte and Majuba Hill; War ends **1881**

Second Boer War begins; Afrikaners besiege Mafeking, Kimberley and Ladysmith; they defeat British field forces at Modder River, Stormberg, Magersfontein and Colenso **1899**

British break Afrikaner sieges; invade South African Republic and Orange Free State; capture Johannesburg and Pretoria. Guerrilla war begins **1900**

Guerrilla war finally ends after British pen most Afrikaner civilians up in concentration camps and crisscross the country with barbed wire **1902**

Battlewagons

Battle of Manila Bay; Battle of Santiago **1898**

Japanese torpedo attack on Russian fleet at Port Arthur; Japanese warships attack Russian ships at Chemulpo (Inch'on) **1904**

Battle of Tsushima Strait **1905**

The Meat Grinder

First Battle of the Marne stops Germans; trench warfare begins **1914**

French offensive, First Battle of Champaign, fails, as does Second Battle of Champaign, Battle of Loos and First, Second and Third Battles of Artois. Germans use gas at Ypres but fail to break through. Gallipoli, in Turkey, is another bloody failure. **1915**

Battle of Verdun; Battle of the Somme; tanks used and misused for the first time **1916**

Massive mutinies in the French army following failed offensive; Battle of Cambrai, first use of massed tanks foiled by sea of mud **1917**

Russia leaves the war; the Ludendorff offensives fail to end the war before the arrival of Americans. Massive allied tank offensive on August 8 was "the black day of the German army," according to General Erich Ludendorff. German navy mutinies; Germany surrenders. **1918**

War Beneath the Waves

Bushnell's *Turtle*, first submarine in combat **1776**

U.S. Navy purchases *Alligator*, a mine-laying submarine **1862**

C.S.S *Hunley* sinks U.S.S. *Housatonic* **1864**

U.S.S *Holland* commissioned **1900**

U-boat sinks three British cruisers **1914**

U-boat sinks two British battleships; U-boat sinks liner *Lusitania*, outraging U.S. public; Germany drops unrestricted submarine warfare **1915**

Germany renews unrestricted submarine warfare; Zimmermann telegram; U.S. declares war on Germany; convoys begin **1917**

Fall of France opens way for German U-boats **1940**

German subs institute "wolf pack" tactics **1941**

U.S. enters war; "happy time" for German subs off U.S. East Coast because of shortage of convoy escorts **1942**

More convoy craft, including aircraft, improved radar and cracking German naval codes make this "the year of the slaughter of the U-boats" **1943**

U.S. submarines finally get good torpedoes and begin destruction of Japanese naval and merchant shipping **1943**

German subs get schnorkels by end of war—too late. U.S. subs by war's end sink 60 percent of all Japanese merchant ships and 201 of the 686 Japanese warships, including the newest and largest aircraft carrier in the world. **1945**

Blitzkrieg and Antiblitz

Germans invade Poland and conquer it in one month **1939**

Germans conquer Denmark and Norway in two months; conquer Netherlands in four days, Belgium in 18 days, drive the British army off the continent in 25 days, and conquer France in 45 days **1940**

Germans conquer Yugoslavia and Greece in 11 days and invade U.S.S.R. **1941**

Germans fail to take Moscow; lose ground during Russian winter; renew offensive, running into superior Russian tanks but drive ahead until bogged down in house-to-house fighting in Stalingrad   **1942**

Russians encircle Germans at Stalingrad, forcing their surrender; superior Soviet armored forces destroy 3,000 German tanks in Battle of Kursk, losses dwindling German war machine cannot replace. U.S. and British forces invade North Africa, Sicily and Italy.   **1943**

Allied forces land in France and push across the continent; Soviet Union continues counter offensive in the east; all belligerents use newly developed armor and anti-armor tactics; liberation of Paris and Battle of the Bulge   **1944**

Germany surrenders   **1945**

The New Queen of the Seas

Attack on Pearl Harbor; Japanese planes sink British battleship *Prince of Wales* and battle cruiser *Repulse*   **1941**

Japanese carrier planes sink British aircraft carrier and two cruisers in harbors of Ceylon; Tokyo raided by U.S. carrier-based planes; Battle of Coral Sea, first naval battle in which ships were entirely out of gun range; Battle of Midway, turning point of the Pacific War, allowing United States to begin "island hopping" in central Pacific   **1942**

Yamamoto Isoroku, top Japanese admiral, killed by carrier planes   **1943**

Battle of Philippine Sea; Battle of Leyte Gulf   **1944**

Okinawa, last and greatest of the island-hopping battles   **1945**

War Against the Home Front

Hitler's aerial "blitz" against England begins   **1940**

British begin "strategic bombing" of Germany   **1941**

Navigational aids and heavy bombers improve British effectiveness; U.S. Army Air Force joins the campaign   **1942**

Schweinfurt raid shows need for fighter escort in daylight; Hamburg raid creates a firestorm, first of many in German cities   **1943**

German night fighters increase Allied losses in raids, but are countered by new fighter escorts, especially the P-51; U.S. forces establish air base on Saipan   **1944**

Bombardment of Japan by land-based bombers begins; atomic bombs dropped   **1945**

High Tech

Gulf War begins with air attack, followed 40 days later by ground
attack, which lasts 200 hours   **1991**

Afghan War: special forces who locate targets for precision missiles give
native anti-Taliban forces the power to topple the Taliban   **2002**

Iraq War: predominately United States force conquers Iraq in three
weeks   **2003**

Low Tech

President George W. Bush proclaims "Mission accomplished" in
May   **2003**

Guerrilla war grows in intensity, especially in Sunni Muslim areas;
Moktada al Sadr instigates uprising of some Shiite Muslims in April;
Iraq gets "sovereignty" June 28; fighting continues—42 Americans
died in June and 54 in July   **2004**

TURNING POINTS
IN
Military History

The Basics: Point and Edge, Arrows and Armor

Prehistoric hand ax found in England. This type of implement is the ancestor of all edged weapons.

Beginnings

AROUSED APES have been seen throwing objects at perceived enemies and even swinging sticks or similar objects as clubs. The first humans may have done the same. It probably didn't take long, however, for someone to discover that the sharp edge of a broken stone can cut and that the sharp point of a stone or of a broken stick or bone can pierce.

From there, it was a short step to purposely smashing a stone to get a sharp edge or to giving a stick a point by burning one end and rubbing the charred section on a rock. The fire also hardened the point, making it a more effective weapon.

Mankind had passed its first military turning point. From there on, people fought like humans, not animals.

Getting the Point

The point and edge were to be the subjects of centuries, even millennia, of development, becoming ever harder, sharper, more durable, and more convenient, but they were the bases of most weapons until the invention of gunpowder. They are the lethal factor in knives, daggers, axes, spears, swords, arrows, and a host of medieval pole arms with strange names. Even in their crudest and most primitive form, they were deadly. Many years ago, archaeologists found the skeleton of an elephant in a bog in Germany. Between its ribs was the wooden point of a spear some prehistoric hunter had fashioned from a yew sapling.[1] The development of what we now call edged weapons was a complicated process—a process that caused a number of turning points in society.

People learned early that the best kind of stone to make a cutting edge was the hard, glassy type like flint, chert, obsidian, or quartz. After a few centuries of pounding these materials in various ways, they learned to shape them by a technique called "pressure flaking."[2] Flint knapping, or producing objects of flint or similar minerals, became a fine art, one that few moderns have mastered. Flint knappers in places as separated as Scandinavia, Egypt, and Mexico produced long, delicate blades for knives, axes and spears that were objects of beauty as well as utility. But flint, obsidian, and the other desirable stones shared one serious drawback: they were brittle. Sometimes, a broken or damaged flint tool or weapon could be refashioned into a smaller one. More often, the only remedy was to chip out a new one.

In some parts of the world, people found copper in pure, metallic form. When a flintsmith began pounding this new type of rock, he found that it didn't chip or break. It just changed its shape. This stuff obviously had possibilities. With what archaeologists call the "chal-colithic" period, the copper and stone age had begun. The next step was a big one: smelting copper ore and casting metallic copper. Some peoples, like the Indians of northern North America, never got that far. The old idea that someone discovered smelting when he used copper ore to make a hearth for his fire probably never happened, since no campfire could produce enough heat to smelt most copper ores. If compounds containing copper are heated to a high enough temperature, they combine with other elements—carbon is the most important one—and free the copper. One ore, malachite, a green carbonate of copper, requires less heat than most.[3] It was often used as a cosmetic. If a lady dropped her flask of eye shadow into a roaring fire, she might later have found a copper bead in the ashes. When people started to settle in permanent villages, they learned how to produce more heat than they could get from a simple wood fire. To make some kinds of pottery, or to bake dozens of loaves of bread at a time, they invented sophisticated ovens that were hot enough to smelt other copper ores. However it happened, somebody did discover how to smelt copper ore, and the knowledge eventually spread all over the Old World. The Indians of Mexico used copper on a small scale. When he marched against Pánfilo de Narváez, who had come to arrest him, Hernán Cortés was accompanied by a contingent of Indian allies armed with copper-tipped pikes.[4] The Peruvian Indians also had copper. Both the Mexicans and the Peruvians, however, relied on stone, rather than copper, points and edges. Copper also had a drawback: it was too soft to take and hold a good edge. Coppersmiths in Europe and Asia learned that they could harden copper to an extent by hammering it, but even that copper left much to be desired.

Flint and obsidian, on the other hand, could be made ferociously sharp. Early in the twentieth century, an American archer, Dr. Saxton T. Pope, showed that an obsidian-tipped arrow penetrated 25 percent farther in hide and meat than a steel-headed arrow of the same weight and length shot from the same bow.[5] But obsidian is brittle. Copper is not brittle, but it can't be made sharp. Something better was needed.

That something was bronze, an alloy of copper and tin. The first bronze may have been made accidentally when some copper ore was mixed with tin ore. Copper is a soft metal, and tin is much softer than copper, but when they're mixed in a molten state, they cool to produce a much harder alloy than either of its constituent metals. Bronze was durable and it would take a pretty good edge. After a while, probably not a long while, the copper smiths realized that a new metal, tin, had been added to their products, and they began to look for tin ore. And that led to the discovery of bronze's main drawback: tin is not nearly as plentiful as copper. The use of bronze stimulated long-distance trade—another turning point.[6] Ships began sailing from the eastern Mediterranean to the far end of the world: the island of Britain, one of the places with an abundance of tin.

The first uses of bronze were for what the kings and chiefs thought were most important: weapons. After military requirements were met, the new metal was used for farming tools, household implements, and such things as mirrors. Bronze made possible a new weapon, the sword, which remained one of the warrior's most important tools right up through the nineteenth century.

Because tin was rare, bronze was expensive. At first, only the powerful could afford it, and the possession of bronze weapons made the powerful even more powerful. In Greece, chiefs of tribes tucked away in that land's many valleys built up their power and created the Mycenaean kingdoms, which were highly organized, hierarchical states ruled by almost-divine monarchs.[7]

That bronze-based social order lasted for several centuries in Greece and other places like Syria and Anatolia. Its end was caused, to a considerable extent, by another metal.

The Iron Revolution

Iron had been used all through the Bronze Age, but seldom for weapons. Its main use was for jewelry, because iron was fabulously expensive. It was, according to some ancient texts, "the copper from heaven." The only source of iron was the nickel-iron found in fallen meteorites. Somewhere in eastern Anatolia or western Iran, ancient centers of bronze making, people learned how to get iron without finding a meteorite. If you took certain rusty-brown rocks, broke them up, and heated them intensely, you would get a mass of rock containing metallic iron. Then you heated the mass some more and pounded it with sledge hammers. The heating and pounding eventually eliminated most of the rock slag. If you did it long enough, you got a piece of almost pure iron—what today we call wrought iron.[8]

Wrought iron was far from the ideal weapon material, but if you heated it some more in a charcoal fire, as you would have to if you wanted to forge it to a useful shape, it would absorb carbon from the charcoal. Iron with a bit of carbon in it is called steel. Smiths learned to control the amount of carbon in various ways. In Japan, they did it by pounding a bar out flat and heating it in the fire, so as much surface as possible was exposed to the charcoal. The Japanese smiths then folded it again, flattened it again, and heated it again. Some Japanese swords are said to have as many as four million layers of steel welded to each other in this folding process.[9] In Europe, India, and Iran, the smiths braided strips of high-carbon iron and low-carbon iron together in a process called "pattern-welding."[10] In India, too, ironworkers heated iron in crucibles with charcoal for prolonged periods and produced small cakes of crucible steel that were much in demand in the Roman Empire.[11] The Chinese invented a blast furnace that let them produce liquid iron direct from the ore long before Europeans could.[12]

Steel development progressed for centuries, but toward the end of the second millennium B.C. it had progressed enough to turn the western part of the civilized world upside down.

This happened for two reasons: weapons of steel were far superior to those of bronze and they were more available. Producing steel from iron ore was a lot more work than producing bronze, but iron ore is found almost everywhere. Copper is not. And tin was much rarer than copper.

For a long time, the ironworkers were under the control of the Hittite empire. The Hittites limited the export of iron objects and guarded the secrets of ironworking. Their iron weapons gave them a military advantage, and they needed any advantage they could get. Their "empire" was really a confederation of independent states, speaking many languages and, according to the Egyptians, worshiping a thousand different gods. They fought the powerful Egyptian empire to their south to a standstill, but at the same time they were being attacked by the hill tribes to their east, the Mycenaean Greeks to their west, and the Phrygians to their northwest. About fifty years after the Greeks destroyed the city of their Trojan allies, the Hittite empire collapsed. Iron weapons and expert iron smelters spread all over the eastern Mediterranean area.[13]

The Aegean area—Greece, western Asia Minor, and the islands between them—was already in something approaching chaos. "The Isles were restless, disturbed among themselves at one and the same time," reads an Egyptian inscription.[14] The availability of iron weapons from Hittite arsenals and newly made weapons by refugee ironworkers—and by their pupils in the lands they had fled to—exploded. Ivor Gray Nixon, using archaeological evidence and the documents surviving this chaotic period, deduces, in his book *The Rise of the Dorians* a tale of turmoil in the Greek world that includes the Trojan War, two attempted invasions of Egypt, and the destruction of almost all the Mycenaean kingdoms.[15] (The Kingdom of Athens was not destroyed, but, like the rest of Greece, it sank into a "Dark Age.")

Much of the *Rise of the Dorians* is conjecture. It is certain, though, that Troy was destroyed, that the Hittite power was extinguished, that Egypt was twice attacked by what the Egyptians called the "Peoples of the Sea," that the Mycenaean kingdoms were destroyed, that this all happened within a hundred years, and that this coincided with the diffusion of iron weapons. The whole known world (i.e., known to the Egyptians, the Greeks, the Hittites, and other Near Eastern peoples) changed. The Egyptians say the "Peoples of the Sea" were joined by the Libyans from the desert west of Egypt and that they included a large number of peoples who have been identified by authorities as Achaean Greeks (Mycenaeans), Dorian Greeks, Philistines, Sardinians, Etruscans, Lycians, Sicilians, and others. It was a mass movement, a wandering of peoples that moved down the east shore of the Mediterranean by land and sea, crushing Anatolian and Syrian states on the way to Egypt. Pharoah Merneptah defeated the first invasion, which included Mycenaean warriors and occurred soon after the destruction of the Hittite empire. The second invasion, a generation later, included Dorians, but not Mycenaeans, and was repelled by Pharoah Rameses III.

After their defeat, the Sea Peoples moved on. Some may have gone back to where they came from, but the Philistines did not return to their Aegean homeland. They settled on the coast of Palestine, where their monopoly of iron weapons made life difficult for other nations, such as the Hebrews. The Etruscans—if that's who the people the Egyptians called "Trshw" were—presumably went to Italy and the "Tshkr" to Sicily. The "Shrdn," or Sardinians, had previously fought for Egypt as mercenaries. Whether their home was Sardinia at that time or whether they conquered it after their defeat at the mouth of the Nile is not known. The Dorians, according to the evidence Nixon musters, went to Crete, conquered it, and then invaded the Mycenaean mainland.

All of this, of course, occurred thousands of years after the first spears, axes, and daggers were produced. In the meantime, another weapon had been invented that marked a major turning point in the military art. And, as it made possible several new tactical systems, it was also responsible for other turning points.

Death at a Distance

The first users of spears and axes were hunters, since agriculture had not yet been invented. Obviously, you can hunt with a spear. People did so for thousands of years. But because the range of a thrown spear is extremely limited, it isn't easy to get close enough to an animal to kill it. Some peoples used a gadget called a spear thrower, which was a stick with a hook or a notch in one end. The hunter fitted the butt of his spear into the hook or notch, held both the spear and the spear thrower in the same hand, and flipped up the end of the thrower as he released the spear. The spear thrower in effect lengthened his arm and gave the spear more speed. The spear thrower is sometimes called the *atlatl*, its Aztec name. The Aztecs used atlatls to throw harpoons, which were used to capture victims to sacrifice to their gods.[16] They believed they had to sacrifice thousands of victims a year to keep the gods happy. An American archaeologist, using an atlatl like those of the ancient Anazazi of the U.S. Southwest, was able to hit a four-inch-square target at forty yards.

The atlatl increased the range of a spear considerably, but for hunting, it didn't compare with the next invention: the bow. The bow was the most efficient missile weapon before guns were invented, and it remained so long after guns arrived. In the eighteenth century, using "composite bows" made of the traditional horn, wood, and sinew, Turkish archers were sending flight arrows up to half a mile.[17] Using fiberglass bows of the type called "compound" bows, modern archers shoot much farther, but the bow is no longer a weapon of war.

It definitely was a weapon of war as well as a hunting weapon in the Stone Age and later. Ten-thousand-year-old rock paintings in Spain show prehistoric men shooting at each other with bows and arrows.[18] Some prehistoric bows from this era have been found in the bogs of northern Europe. They are remarkably like the famous English long-bow that marked another turning point millennia later. They are what are called "self bows," bows made of a single type of material, in these cases, wood. Wooden self bows are not as flexible as composite bows, so they have to be made longer to draw an arrow of the same length efficiently. Length has nothing to do with the strength of the bow, although most of the writing about the longbow implies that it does.

A longbow could shoot a heavy war arrow about 250 yards, which made it, at a time when game was plentiful, an extremely efficient hunting tool.[19] It was also the prime war weapon for hunters, herdsmen, and others living in the wide open spaces who needed something that could be used on either animals or humans.

People who lived in towns and cities didn't get as much practice with the bow. For most of them, the main weapon was the spear or the ax. They had to get close to their enemies to fight, and they had to have some way to get close without being pincushioned with arrows.

They invented armor.

The earliest forms of armor had little or no metal. Quilted coats or jackets made of many layers of linen were an ancient kind of armor that remained popular for a long time. Greeks in the classical period wore linen corselets, and Aztecs in the sixteenth century had quilted cotton jackets.[20] Their Spanish opponents found these jackets better than their own steel armor for strenuous campaigning in the mountains of Mexico. The "Royal Standard of Ur" an ancient Sumerian panel with inlaid mother-of-pearl figures on both sides, shows Sumerian warriors with another kind of primitive armor: copper helmets and long leather cloaks studded with metal disks.[21] The most effective armor, though, was the shield. If body armor was pierced, the wearer was wounded. But a shield was held away from the body. To damage the user, the arrow had to penetrate the shield for at least half the length of the arrow, and then the almost-exhausted arrow had to pierce the shield-user's body armor.

Most early archers didn't wear armor because it interfered with their freedom of movement and mobility. And it is extremely difficult to shoot a bow and hold a shield at the same time. City-dwellers, though, made good use of armor, and they developed a couple of formations to make the most of it.

2

Urbanite Warfare

Greek soldier in full armor. Masses of hoplites like this
made up the ancient Greek phalanx.

Marathon

THERE WAS A SWARM of Persians on the plain of Marathon, so many that Miltiades, the commanding Athenian general on this day, found it advisable to lengthen his line.[1] The traditional Greek formation, the phalanx, consisted of eight ranks of armored spearmen, a formation that did not vary from city-state to city-state. Miltiades reduced the number of ranks in the center of his army from eight to four, giving him a longer line. The ends of the line kept the traditional eight ranks.

It wasn't the Persians' sheer numbers that bothered Miltiades. Miltiades' army consisted of 10,000 Athenians and 1,000 Plateans. The size of the Persian army was grotesquely inflated by Greek historians of a later generation. In spite of them, there certainly were not 100,000 Persians on the plain. Herodotus gives a realistic account. He says the Persians arrived in 600 ships.[2] Admiral William L. Rodgers, in his authoritative *Greek and Roman Naval Warfare*, estimates that the Persian fleet could have carried thirty soldiers per ship or five cavalry soldiers and their horses per ship. He believes the fighting force consisted of 6,000 marines, who were normally stationed aboard ship for boarding and similar operations, 1,500 cavalry, and 7,500 infantry—a total of about 15,000 combatants.[3]

Insead of numbers, Miltiades was concerned about the Persians' mobility. The Persian force was a combination of cavalry and light infantry, most of them archers. The Athenian general had no wish to see these "barbarians" flowing around his flanks and falling on his men from the rear. He lengthened his line so there wouldn't be enough room on level ground for enemy flanking maneuvers.

The Greek army held the mountains ringing the plain of Marathon. The Persian army wasn't big enough—especially since so much of it was cavalry, which is not adapted to mountain warfare—to successfully attack the Greeks. Then someone saw what looked like a signal: sunlight flashed from a polished shield toward the Persians. The Persians then began loading their ships.

When the Persians started embarking their horses, Miltiades understood that the landing at Marathon was a feint to draw the

Athenian troops out of Athens.[4] A faction in Athens had agreed to open the gates of the city to the Persians if the invaders could get the Athenian army far enough away. The seaborne Persians hoped to sweep along the coast and get to Athens before the heavily armed infantry could march back over the mountains to their city. Meanwhile, the Athenians had been waiting for reinforcements from Sparta. They had no doubt that with such reinforcements, they could crush the Persians. But now there was no time to waste. It was either attack now or risk losing their home.

Miltiades gave an order, and trumpets sounded the charge. The musicians played a marching tune on their flutes, and the Greek army set out down the slopes toward the Persians on the plain. They marched slowly in step, in time with the music. Marching in step was a Greek invention. It let the phalanx march like a giant machine, a machine coated with bronze armor and bristling with iron-tipped spears. Each *phalangite* supplied his own armor, but all of it was similar. Most of the bronze helmets covered the soldiers' heads entirely except for the eyes and mouth. The troops carried round, bronze-faced shields, three to four feet in diameter, that covered almost all of their torsos, exposing only enough of the right side to let them thrust with their spears. Each man crowded close to the soldier on his right to get the added protection of that man's shield. The men's legs below the shields were shielded with bronze greaves.

When the Greeks were about 200 yards away, the Persian archers began shooting. At that range, the arrows merely bounced off the oncoming bronze glacier. The Greek trumpets gave another signal, and the slow march changed to double-time. The glacier turned into an avalanche, but an avalanche that rushed forward in step. For several centuries, the Greek heavy infantry had worn a solid bronze corselet that made running difficult. But in recent years, the Greeks had exchanged the solid corselets for more flexible body armor: corselets made of many layers of linen or fabric covered with metal scales.[5] At close range, the Persian arrows could penetrate armor, but they glanced off the curved bronze helmets and greaves. The armor most of them hit were the shields. The Greeks, of course, held their shields away from their bodies, so the arrows—even if they penetrated the shields entirely—never hit a vital spot. The shield was the most important part of a phalangite's equipment. The Greeks called their heavy infantryman a *hoplite*, the soldier who carried the *hoplon*, or shield.[6]

When they got close to their enemies, the Greeks thrust overhand with their spears, which were longer than those of the Persians. Many of the Persians had no spears because they relied on their bows, and few of them had armor. The heavy Greek wings drove into their foes. The Persian flanks were held by troops from the subject nations, neither well armored nor strongly motivated. They were forced back. In the center, it was a different story. That part of the Persian line was held by ethnic Persians and savage Scythian nomads. These warriors fought fanatically against the weakened Greek center, even scrambling over the wall of shields to hack at the phalangites with dagger and battle ax. The Greek center fell back under the pressure of the Persians. The wings, pushing ahead and trying to keep in contact with the center at the same time, executed what amounted to a pincer movement. When the Persians saw that they were about to be surrounded, they turned and ran. The whole Persian army dashed to their ships.

The Greek phalanx had won what most authorities consider one of the most decisive battles in history. Most of the Persian army escaped, but the Athenians at Marathon sent a professional runner back to Athens with word of their victory. The city's gates remained closed while the army made a forced march over the mountains.

The Phalanx and Democracy

The Greeks didn't invent the phalanx. That honor probably belongs to the Sumerians, many centuries before Marathon. The "Royal Standard of Ur," a panel covered with figures carved from shell and limestone and found in a plundered grave in the royal cemetery of the ancient city of Ur, shows a primitive form of phalanx. Soldiers bearing short spears and wearing tight-fitting copper helmets and long leather cloaks covered with metal disks march in line.[7] They have no shields, but another Sumerian sculpture shows a mass of spearmen lined up behind a wall of big oblong shields.[8]

The Sumerians were among the earliest city-dwellers and the first literate people. They built towering step pyramids crowned with temples, constructed massive walls around their cities, and dug a maze of irrigation canals to use the waters of the Tigris and Euphrates to grow their crops. They were used to cooperating with each other. The phalanx was based on cooperation—a mass of spearmen moving as a single unit to get the maximum shock effect. It was a natural way for city-dwellers to fight.

What the Greeks did was perfect the phalanx. They introduced marching in step and trumpet signals. And they protected their hoplites with extremely effective armor. The prime time for campaigning was early summer: the crops had been planted and they didn't have to be harvested right away. Early summer in Egypt and Mesopotamia is furnace-hot, which may be why these ancient centers of civilization did not emphasize metal armor. It can be pretty warm in Greece, too, but Greek armies seldom marched very far from their home cities. And when they did, slaves carried their armor. The phalanx was ideal for fighting in the narrow valleys of that mountainous country, and the Greeks got plenty of practice using it. Usually, in any given year, several of the Greek city-states were warring against each other. And all that phalangial fighting led to an idea that would change the world.

Democracy.

The first Greeks were the Mycenaeans. From what historians can gather from their inscriptions, they were anything but democratic. Each city was led by a king, called a *wanax*, who was a link between his people and the gods. He lived in a palace and headed an elaborate bureaucracy, which administered everything from food production to armaments. Warships and chariots were important parts of Mycenaean military machines.[9] The chariots were probably ridden by aristocrats, some of whom wore extremely sophisticated, but heavy, bronze armor.[10] Almost all of the Greek myths and legends we know, including Homer's tales of the Trojan War, come from the Mycenaean period. In describing the tactics of the Trojan War, Homer probably included details from his own, prephalanx period, but it seems clear that the main emphasis in Mycenaean warfare was combat between individual champions, usually kings and princes. They rode chariots, but usually dismounted to fight. They were extremely well armed, but their followers were not, since bronze was so expensive.

The beginning of the iron age saw the fall of the Hittite empire, the invasions by the "People of the Sea," and a long period of civil war in Greece. These events were followed by what historians call the "Dark Age." Writing apparently died out and archaeological digs indicate deep poverty for several centuries. Cities were rebuilt and petty kingdoms were established. And, it seems, the phalanx was born. At first, the phalangites came from a city's aristocrats, the same class that furnished the armored chariot warriors in Mycenaean days. Chariots were out. They were expensive and really weren't of much use in a land of rugged hills and narrow valleys. Slowly, seaborne trade was revived, and the Greeks adopted a new form of writing, the alphabet, from the Phoenicians. With revival of trade, more and more Greeks were able to afford bronze armor. In phalangial warfare, numbers were extremely important. It didn't pay to exclude newly rich citizens from the phalanx just because they weren't aristocrats.[11]

A word about bronze armor: Iron was more plentiful than bronze, but bronze was the Greeks' choice for armor because of technical reasons. The furnaces of the time were hot enough to melt bronze, but not iron. It was easy to cast bronze into an object of any size or shape. On the other hand, iron could be extracted from the ore only in relatively small pieces. To make an object the size of a sword, a smith had to weld many pieces together. It just wasn't practical to weld and temper enough bits of iron together to make a solid iron breastplate.

Bringing commoners into the phalanx was a step toward democracy. More important, though, was the way a phalanx fought. There was no room for heroes leading a rabble of followers from their chariots. The Athenians had cavalry—rich men who could afford horses to ride to the scene of a battle. But when a fight was about to start, these horsemen got off their mounts and took their places in the phalanx. Even the commanding generals fought as simple hoplites. Aside from ordering advances and retreats, there was little a commander could do except take his place in the lineup and poke his spear at the enemy.

Kings, originally thought to have a direct line to the gods, were supposed to lead their people in battle, but you didn't need divine inspiration to lead a phalanx. So most cities got rid of their kings. The conservative Spartans kept not one, but two, kings to provide checks on each other, and then they deprived them of almost all power. Of course, Greek democracy was not all sweetness and light. From time to time, a "strong man," something like a modern mob boss, would appear and take power in a city. He was called a tyrant, a self-made king with no traditional basis for his authority. And after a while, the citizens would overthrow him. Some cities, like Athens, adopted antityrant laws.

After the Persians were finally repelled, Greece entered the "Great Wars" period. The first was the Peloponnesian Wars, a fifty-six-year-long affair between Athens and Sparta for the leadership of Greece. It was followed by the Corinthian War, then by years of fighting between Athens, Sparta, and Thebes, and was occasionally mediated by the king of Persia. For centuries, the phalanx had been sufficient for the continual, but limited warfare between the Greek cities. But in the Great Wars, fighting was no longer limited. Combat was no longer a kind of deadly game played on the nearest level ground, with the defeated admitting they had lost and asking permission of the victors to remove their dead. Now, fighting took place in the mountains, on small islands, and on every kind of unfavorable terrain. Greek generals began inventing new tactics.[12]

Iphicrates, an Athenian, raised a body of mercenary light infantry, called *peltasts* from the type of shield they carried. The peltasts were javelin-throwing, highly trained, professional fighting men, not citizen soldiers. Iphicrates' force of peltasts could move easily on broken ground, unlike the unwieldy phalanx. The peltasts could stay out of reach of the hoplites' spears, surround them, and attack from all sides. Greek commanders began augmenting their forces with professional slingers and archers, whose weapons had even more range than the javelins. They even brought in javelin-throwing horsemen from Thessaly.

One general, the Theban Epaminondas, continued to rely on the phalanx, but his was a different type of phalanx. Epaminondas noticed that as hoplites marched, they unconsciously moved obliquely to the right, as each man was trying to get behind part of the shield of the man on his right. The Spartans noticed this, too. They put their best men on the right of their own line so that they would overlap the left of their enemies' line and, when contact was made, attack the enemy's left flank and roll up his line. Epaminondas put his best men on the *left* of his line, lining them up in a pile driver of a phalanx— fifty ranks deep. Then, like a good boxer, he led with his left and held back his right. The Theban army marched in an oblique order and consciously moved a little to its left as it marched. The Spartan right wing did not overlap anything, and the left was still far from the enemy when the right was engaged with the Thebans at Leuctra. Epaminondas's pile driver drove back the Spartan right. The Spartans ran—something the Greek world considered impossible. Sparta, which had been considered practically invincible in shield-to-shield land combat, never fully recovered from Leuctra.

The Macedonian Machine

A young hostage from the semibarbarous kingdom of Macedon was with the Thebans and watched Epaminondas crush the Spartans. He also had an opportunity to observe how the other Greek cities used light troops. When he went home, the young man who became Philip II of Macedon, took the lessons he learned, added many original ideas of his own, and created a new kind of army. It is known to history as the Macedonian Phalanx, but the actual phalanx was not the most important part of it under Philip and his son, Alexander the Great.

The Macedonians were Greek hillbillies who still had a king and nobles. And the nobles were all horsemen. Not just men who rode horses, but men who began riding horses soon after they were able to walk. Even without stirrups, which would not appear in Europe for many centuries, they could maintain a firm enough seat on their horses to thrust with lances while wearing heavy armor. Philip added a heavy cavalry unit, called the King's Companions, to his army. He also had light cavalry armed with javelins, and light infantry using javelins, bows, or slings. Philip introduced a brand-new arm: field artillery. He used torsion-powered catapults for both long-range fight-ing and sieges.[13] And he remade the old phalanx into something both more massive and more mobile than the old Greek model.

Philip's phalanx was sixteen, instead of eight, ranks deep, but the phalangites were not as heavily armored as the Greek hoplites. They didn't wear corselets, and their shields were smaller and worn on the left arm, but without a handhold. The Macedonian phalangite needed both hands to manage the *sarissa*, the spear Philip introduced. It was at least ten feet longer than the eight-foot hoplite spear. In battle, the first five ranks leveled their spears. An opposing hoplite would have to make his way through this wall of points before he was close enough to use his own spear. Philip did not use his phalanx, as Epaminondas did, to concentrate power on the decisive point. He reserved that role for his heavy cavalry. Connecting the heavy cavalry and the phalanx was a new kind of infantry, the *hypaspists*. They wore armor heavier than the phalangites' but lighter than the hoplites' and carried spears like the old hoplite model. They could fight as a phalanx or in extended order.[14] The basic tactic of both Philip and Alexander was to hold the enemy with the phalanx and strike him with the cavalry. Or as General George S. Patton put it, "Hold him by the nose and kick him in the ass."

The Macedonian machine was an extremely sophisticated mili-tary instrument. But it took an extremely sophisticated general to get the most from it. Philip was such a general. So was his son, Alexander. But after Alexander's death, his heirs opted for weight instead of mobility. The phalanx kept its long spears, but it got heavier armor. The proportion of cavalry to infantry dropped sharply and the Macedonian generals introduced an exotic beast from the East: the elephant.

The Roman Legion

The changes were all in the wrong direction, as the Greeks and Mace-donians realized when they confronted a new infantry formation: the Roman legion.

Centuries before, the Romans had adopted the Greek phalanx. As with the Greeks, the phalanx led to democracy of a sort. The Romans had five classes of soldier, armed according to their wealth. Men in the first class were armed the same way as the Greek hoplites. Second-class soldiers had bronze breast plates, helmets, shields, and greaves. Men in the third class had the same, but no greaves or breast plates. Fourth-class troops had shields, leather jackets, and helmets of leather or pottery. The fifth class had only their weapons. The first three classes of legionaries used a short sword and a long spear, the fourth class had a sword and javelins, and the fifth class had javelins or slings. In combat, the first three classes fought as a phalanx. The third and fourth classes were light infantry who opened the battle with their missiles and then retired behind the phalanx.[15]

Fighting against the Samnites, hill tribes who fought in a loose formation, and the Gauls, who relied heavily on cavalry and mounted infantry, the Romans found their phalanx too slow and inflexible. Marcus Furius Camillus took the first steps to modernize the legion. He replaced the phalanx with lines of *maniples* (Latin for handfuls). There were 120 men in a maniple, commanded by two centurions. They stood in a block twelve men wide and ten ranks deep. In battle formation, the legion had two lines of maniples arranged like the squares on a checkerboard. The maniples of the second line covered the intervals in the first line. The third line of Camillus's legion was a phalanx of men called *triarii*. They carried long spears. The two lines of maniples in front of them carried two throwing spears, a light one to throw when within twenty yards of the enemy and a heavy one to throw just before making contact. All of the legionaries carried a sword.[16] To open the battle, Camillus used light troops—javelin men and slingers.

The Roman throwing spears, called *pila*, were designed to bend if they stuck in an enemy's shield, so they could not be pulled out and thrown back. When the legionary reached his enemy, he'd step on the shaft of the dragging spear, forcing the enemy's shield down, and then attack with his sword.

Enemy phalangites were apt to flow around the first line of maniples, thus losing the cohesion that was the phalanx's greatest strength and leaving them vulnerable to attack by the second line of maniples and by the Roman third line, the legion's own phalanx. And in really difficult terrain, the Roman maniples had no trouble while a phalanx was stymied. That prompted the Roman generals to break the triarii into maniples, too. The commander of a legion could always form a phalanx just by ordering his maniples to close up, as Varro did when he faced Hannibal at Cannae, but he could also fight in open order. Scipio Africanus trained his legionaires in more sophisticated maneuvers so he could concentrate his forces at any point in the line.

Most of Rome's enemies used a version of the phalanx. Hannibal was an exception. In Italy, his army was so heterogeneous that such an organization was almost impossible even if the commander wanted it. And he didn't. All the Hellenic heirs of Alexander used it, but they had transformed the agile Macedonian machine into something slower and more hide-bound. Neither Alexander's heirs nor the Romans had anything like Alexander's powerful cavalry. The Romans used cavalry from their allies or conquered nations, but it wasn't the same.

Gaius Marius reorganized the legion by concentrating the maniples into cohorts of about 600 men each, which still fought in three lines. He also instituted a professional army, replacing the old array of amateur citizen-soldiers. This proved to be a step away from democracy, but it made for a more efficient army. The Roman legion was the most successful infantry formation before the invention of gunpowder. But it didn't conquer the known world because at the gates of Persia it ran into another successful and more ancient type fighting force.

3

The Rulers of the Plains

Central Asian horse archers. Warriors like this were practically invincible on the steppes until guns were introduced.

To Conquer the East

WHEN THE ROMAN legions—40,000 infantry accompanied by 4,000 allied cavalrymen—splashed across the Euphrates, many of them were sure they were beginning a great adventure. None were surer than their commander, Marcus Licinius Crassus. Crassus, a shrewd soldier-politician who had put down the Spartacus rebellion, was one of the three most important men in Rome, and the richest of the three. Each member of the triumvirate was busily trying to undercut the other two. The three, Crassus, Gnaeus Magnus Pompeius (commonly called Pompey), and Gaius Julius Caesar, had agreed to divide up the Roman Empire into spheres of influence. Pompey got Spain, Caesar got Gaul, and Crassus got Syria. Spain had silver mines and controlled the Pillars of Hercules—the gateway to the great ocean and all the lands it touched, but there was no evidence that there was any land in those western seas worth having. Pompey was currently the big man in Rome, but that was bound to change. Caesar had all Gaul, but the Gauls were semibarbarians, and their only neighbors, aside from Italians and the Spaniards, were the Britons, who were their cultural equals, and the Germans, who were pure barbarians. Syria was different. It was civilized long before Rome itself. More important, it was the gateway to the East.

The East was the home of fabulously rich kingdoms. By crossing the Euphrates, Crassus had entered the first of them—Parthia, the successor to the ancient empire of Persia. Beyond Parthia was India, and beyond that, the legendary empire of China. Almost three centuries before this, Alexander the Great had led his phalanx into India. If his men hadn't become homesick, he would have continued to the eastern end of the world. Crassus believed Romans were made of sterner stuff than Macedonians and Greeks. He would conquer China and return with so much wealth and prestige Caesar and Pompey would have to give way.

Crassus didn't realize that the Persian Empire Alexander had conquered had been weakened by rivalries among the royal family, corrupt officials, and ethnic enmities among its population. Its army was a hodgepodge of troops from all over the vast empire, all using different weapons and tactics and commanded by a king who was a military moron. After Alexander won his first three battles, most of the large cities he came to threw their gates open. For a while, the Greek Seleucids who inherited a portion of Alexander's domain, ruled most of the old Persian Empire. But that was before the Parthians arrived.

The Parthians had been nomads on the Central Asian steppes. Nomads herding cattle and horses had to be mobile, so all the Parthians were horsemen. Their herds had to be protected from predators—

The Composite Bow

The composite bow invented by Central Asian nomads compared with the famous English longbow the way a modern rifle compares with a flintlock musket. The bow was composed of a thin piece of wood that served as a foundation for the elements that provided the power. On the back of the bow, the nomads glued a thick layer of sinew, which could stretch and then snap back rapidly. On the belly of the bow (the part facing the archer when he's shooting), they glued a layer of horn, which can be compressed but powerfully reasserts itself.

It is generally conceded that the Turks made the best of all composite bows. In the late eighteenth century, the longbow was making a comeback in England—as a sporting weapon, not a tool of war. In 1795, a Turkish diplomat was invited to show some leading English archers what a composite bow could do. He shot an arrow 482 yards, which boggled the minds of the English, whose longbows could seldom exceed 250 yards. The Turk, though, apologized, explaining that both he and his bow were stiff because of a long period of inactivity. He said his monarch, the sultan, could shoot an arrow 800 yards.

wolves, leopards, and even the occasional lion or tiger. The predators were fast and the herds were dispersed over the plains, so the nomads needed a weapon that was effective at a distance. They developed a short but powerful bow that could be used on horseback. It was composed of horn, wood, and sinew, instead of simply wood. All the steppe nomads were mounted bowmen. They all used the same tactics when they fought over pastures. Nomad wars are history's best example of the "domino effect," which was heavily discussed during the Cold War. The losers of these wars pushed into the territory of a weaker nation, and the loser of that conflict moved against still another tribe. Or into the settled world.[1]

It was probably pressure from other nomads that brought the Parthians into the crumbling kingdom of the Seleucids. They spoke a language similar to that of the Persians and quickly adopted some Persian institutions. The most important military feature was armored cavalry. The wealthier Parthians started wearing armor composed of iron strips laced together and began using lances as well as bows. The Greeks fought them for seven years, and for a while the Parthians had to submit. A little more than a century before Crassus entered their territory, the Parthians revolted and drove the Greeks across the Euphrates.

The Parthian king was worried when he heard that the powerful Romans were invading his land.[2] He wasn't anxious to face them himself, so he asked his strongest vassal, the king of Suren, to take care of them. If the vassal won, all would be well. If he didn't, the great king would at least lose his greatest potential rival, and the Romans would be weakened. The vassal, called Surena by the Romans, was only thirty. The Romans said he painted his face like a girl and took a bevy of concubines on campaign with him. But among the Parthians, Surena was famous for his valor. He soon would be for his sagacity. He needed both courage and wisdom, for he had only 10,000 cavalry, 1,000 of them armored lancers, to meet the advancing Romans.

The invasion at first went swimmingly from the Roman point of view. Parthian light cavalry shot at them from a distance, but retreated continually. Then an Arab who had given the Romans intelligence in a previous campaign showed up with more information. He said the Parthian king, Hyrodes, was hiding in the wilderness and sent Surena to divert the Romans with a small force. The best thing for the Romans to do, he said, was to quickly crush Surena's army and then go after Hyrodes. If they went on as they were doing, they'd never catch Surena's fast-moving light cavalry. But, the Arab said, he knew a shortcut that would let them take Surena by surprise. Crassus ordered the Arab to guide them on the shortcut.

Following the Arab, the Romans left the verdant river valley and headed into the desert. They marched for miles across the bleakest terrain any of them could imagine, trudging through soft sand under a blazing sun with no water in sight. Finally, they saw a small stream and along its banks, the Parthian army. Surena's army was even smaller than they expected. The Arab said he'd make their victory even easier. Armed with a plan to disorder the Parthian force, he left to put his plan into effect.

Surena was not going to be surprised. He had the bulk of his force hidden behind sand dunes, their armor covered with skins so the Romans wouldn't see the sun glitter on the polished metal. Without warning, the Romans heard the boom of thousands of kettle drums, and the Parthian army suddenly appeared on the crests of the dunes. The Parthian heavy cavalry charged with leveled lances, but the legionaries stood firm. Roman legions would not be moved by mere cavalry. The Parthians turned and fled. They fled in all directions. Then the Romans noticed that the fleeing Parthians had stopped in positions that completely surrounded the Roman army. They began riding around the Romans and shooting at them. Their bows out-ranged any weapons the Romans had, and their arrows could pene-trate the iron Roman corselets. The Romans waited for the Parthians to run out of arrows. They didn't. Surena had brought along a thou-sand camels loaded with extra arrows.

Crassus called his son, Publius, who had served with distinction under Caesar in Gaul, and told him to counterattack. Publius had enlisted a large number of Gaulish horsemen—the best in the West—to serve with his father in the East. He took 1,300 horsemen, 500 archers, and some 4,000 infantry and charged the Parthians. The Parthians fled, shooting back as they ran. When the charging Romans were too far from the main body to receive assistance, the Parthian heavy cavalry charged the Romans. Publius fell back to a defensive position, his cavalry sheltering behind the immovable Roman infantry. The Parthian lancers stopped, but their light horse archers surrounded the Romans and shot them all down. They cut off the head of Publius and threw it into the Roman main body.

The Romans tried to retreat, but they became separated in the dark. The Parthians surrounded the detachment Crassus was com-manding. Some Arabs approached and said Surena wanted to talk with

Crassus about surrendering. But while Crassus and his top officers were discussing terms, a fight broke out and Crassus was killed. The surviving Romans were enslaved.

The Nomads' Ultimate Weapon: The Horse Archer

Crassus was by no means the greatest Roman general, but a much better general, Pompey, was stymied a few years later by the Parthian horse archers. Crassus, Pompey, and every other general who crossed swords with the Central Asian horse archers—right up to the crusaders a thousand years later—met nothing but defeat. Centuries before, Darius the Great lost an army when he invaded the lands of the Scythian horse archers. Alexander the Great did manage a short-lived conquest. Alexander was a genius, of course, but he was also very lucky.

In the Near East and Central Asia, the mounted archer would reign supreme until the coming of gunpowder. And the nomads, masters of mobile warfare, had ruled the plains for centuries before they even learned to ride horses. Their military tradition began when they adopted the chariot. The defeat of Crassus emphasized a turning point that had occurred more than a thousand years before, but was unknown in the West. And it shows how silly are the statements by some modern historians that the West has dominated warfare for 2,500 years,[3] that Western discipline, organization, and ruthlessness had invariably conquered non-western armies,[4] and that Western armies fighting non-Westerners were almost always outnumbered.[5] The last point is particularly ludicrous when you consider what 10,000 Parthians did to 44,000 Romans.

The first recorded contact of the West with the new military machine from the steppes is when the Egyptians of the Middle Kingdom met the Hyksos, the "Lords of the Uplands" or the "Shepherd Kings," around 1800 B.C. We know the Hyksos had chariots and compound bows, because we know that the Egyptians learned how to make and use these weapons from the Hyksos and used them to drive the Hyksos out of Egypt. According to tradition, the Hyksos took Lower Egypt "without a fight." That could mean without a stand-up infantry fight—the only kind the Egyptians were accustomed to. Hyksos charioteers could literally ride circles around the Egyptian infantry. Their chariotborne archers could concentrate their fire on any part of the Egyptian phalanx, and with their composite bows, they could do it while staying out of range of the weaker and far less numerous Egyptian bows.

The Hyksos didn't invent the chariot, of course. The earliest representations of chariots are Sumerian, centuries before the Hyksos even appeared on the scene. These chariots, shown on the "Royal Standard of Ur," are clumsy vehicles with four solid wheels and drawn by a pair of onagers (Asian wild asses).[6] The horse was unknown, and so were spoked wheels. Each chariot had a driver and a warrior armed with javelins. Whether this design somehow reached the steppe-dwellers, or whether they invented the chariot independently is unknown. It's pretty certain, though, that the steppe people were first to domesticate horses.[7] What they did with the horses was to hitch them to a very light vehicle, big enough for two standing men, with a pair of independently rotating wheels. Horses in those days were too small to ride. The nomads tried to breed bigger horses, and after a few centuries they succeeded. In the meantime, their light chariots gave them more mobility than any other people on Earth. A little after the second millennium B.C., chariots suddenly exploded in the East, the West, and the South. Chariots appeared in Iran, India, the Near East, and Europe. In Europe, a mountainous projection of the Eurasian continent, full of forests and swamps, their usefulness in war was somewhat limited. But in northern China and the deserts of the Near East, the Middle East, and South Asia the chariot was for centuries the ultimate weapon. The chariot usually had a two-man crew—a driver and an archer. It could be used as a shock weapon: untrained infantry usually ran when faced with two charging horses. Foot soldiers soon learned, though, that horses would seldom try to break through a line of spear points. But as a means of concentrating firepower on any part of a battlefield, even on the enemy's rear, the chariot had no equal.

In China, chariot warriors apparently founded the first dynasties of which we have records. In what was to become Persia, chariot-riding Iranian tribes, the so-called Aryans, galloped across the deserts and steppes to overcome the infantry of the inhabitants and founded kingdoms that would later unite into the mighty Persian Empire. In India, other Aryan tribes would wipe out the ancient Indus civilization, then pour into the rest of the subcontinent. Other Indo-European-speaking peoples would ride into the Near East on two wheels. The natives there quickly adopted the chariot and pressed on to new conquests.

The Hyksos beat the Egyptians, but the Egyptians then became charioteers themselves and not only drove out the Hyksos but pushed on into the Hyksos homeland, Syria. There, they encountered two other chariot-using nations: the Mitannians and the Hittites. Both the Mitannians and Hittites spoke an Indo-European language. The Mitannians had international fame as horse trainers. A letter from a Hittite king to a Mitannian horse trainer asking for help in this area exists.[8] The Hittites, unlike most peoples, used a chariot carrying three men instead of two, thus increasing their firepower.[9]

Civilized and semicivilized nations all adopted the chariot, even in places like Greece that were not good chariot country. When Julius Caesar went to Britain, he found the native chiefs using an open-fronted chariot that allowed them to run along the yoke pole between the horses to throw their spears. The high-ranking British charioteers loved it, because it let them show off their athletic abilities, though it wasn't very effective.

Horses

In the most efficient armies, the chariot had long before been displaced by cavalry—another nomad invention. One man on a horse, armed with a bow and arrows, along with a battle ax or sword, could travel faster, maneuver more quickly, and traverse terrain completely impossible for a chariot. Nomads were usually horse archers, but some semiurbanized nations like the Macedonians and Persians learned to use the lance even before stirrups were invented. Stirrups, another nomad invention, let a rider stand in the saddle, making it easier to shoot.[10] But most important, they made it possible to concentrate both the strength of the rider and the horse into the point of a lance. The rider could strike a mighty blow without being pushed off his mount.

In the wide open steppes of Central Asia, the Near East, and eastern Europe, and on the Hungarian meadows of central Europe, horse archers were practically irresistible. The Cimmerians, the Scythians, and the Sarmations owned that territory. Some believe the Cimmerians combined with the natives of central Europe to form the people later known as the Celts.[11] Celtic cavalry rode through Gaul and Spain, ravaged Italy and Greece, and settled in Anatolia and Britain. Celtic cavalry tactics differed from those of the nomads. The Celts, like the Romans and Greeks, relied on shock tactics. Their cavalry's main weapons were not bows, but long swords. The Goths and other mounted German tribes who fought the Roman Empire also adopted shock tactics. For them, the lance and the sword, not bows and arrows, were the weapons of gentlemen. Even though western Europe—particularly Italy and Greece—was hardly ideal cavalry country, cavalries gave the city-dwellers in those areas a most difficult time.

The German tribes that adopted cavalry became the most formidable of Rome's foes until the Hunnish horse archers conquered them. When it came to cavalrymen, the horse archer still outclassed the lancer, even in unsuitable country like western Europe. The Roman general Aëtius, a onetime friend of Attila, the Hunnish king, used Hunnish mercenary horse archers to eliminate German troublemakers.[12] When Aëtius defeated Attila at Chalons, it was because Attila had incorporated so many German horsemen into his army that he couldn't use the Huns' traditional horse-archer tactics.[13] Attila's death and the civil war that followed it ended the Hunnish empire, but Asian horse archers raided Europe in a seemingly endless series of waves. The Bulgars, Avars, Turks, Magyars, and Mongols were to terrify Europeans for a thousand years. Some of them raided deep into France and Germany. But the most formidable of all these nomads, the Mongols, after wiping out the armies of the Teutonic Knights, Poland and Hungary, left western Europe alone. The Westerners had introduced something new and brought about another turning point.

4

Stonewalling the Nomads

Assault on a castle. Few nomads could employ battering rams like this. Even with engines, assaults seldom succeeded.

The Riders from Hell

THEY APPEARED in 1222 as suddenly as if they popped out of the ground. Indeed, some Europeans were inclined to think that they had. The Europeans heard that some of the strange horsemen were called Tatars. The name, as used by Europeans, quickly changed to Tartars—inhabitants of Tartarus, the nether region. In a letter to Pope Honorius III, Queen Rusudan of Georgia explained why her Caucasian kingdom would not be able to send an army to join the Crusades, "A savage people of Tartars, hellish in aspect, as voracious as wolves in their hunger for spoils and as brave as lions, have invaded my country."[1]

When the invaders confronted the chivalry of Georgia, they fled into a narrow mountain pass. The Georgians pursued, but when they emerged from the other end of the pass, a second group of invaders attacked them from the rear with volleys of arrows. The first group then wheeled around and attacked their pursuers. After shooting down the majority of the Georgians, the invaders finished off Georgia's defenders with lance and sword. They burned a few villages and massacred the villagers, but they didn't tarry in Georgia. They crossed the Caucasus and headed north.

On the plain below Europe's highest mountains, they saw an army of native mountaineers, Alans and Circassians, and a tribe of Kipchak Turks, the Kumans. The invaders sent a delegation to the Kumans bearing gifts. The delegates told the Kumans that it was not right that they should fight, as they were both nomads of the same blood. The Kumans left the mountaineers and straggled off across the steppe. The invaders then defeated the Alans and Circassians. After a short pause, they attacked the scattered Kuman clans, defeated them, and took back the gifts. Many of the Kumans sought refuge in the Byzantine Empire and brought more news of the invaders. The invaders continued west and entered the Crimea, where they stormed the Genoese fortress of Sudak. The Genoese fled in their galleys and gave the West still more information about the hellish riders. Meanwhile, the invaders rode north. There they saw a huge army of Russians and Kuman survivors that outnumbered them almost three to one. They retreated for nine days, allowing the quarreling armies of the Russian princes to disperse. On the tenth day, they attacked the individual units. About 90 percent of the Russians perished on the field. The strange horsemen then turned east.

Europe had seen for the first time the Mongols of the Gobi. It was not a full-scale invasion, but a reconnaissance in force, although it was led by two of the three greatest cavalry generals in history: Subotai Bahadur and Chepé Noyon. The third general was their master, Genghis Khan, who was now in the Gobi. Subotai and Chepé had not intended to conquer Europe. They had only three divisions of the Mongol army (the commander of the third division had displeased Genghis and had been demoted to the rank of private soldier). Mongol divisions, called *tumans*, had 10,000 men each. A mere 30,000 men could not conquer a continent. But Subotai and Chepé had collected an enormous amount of information, from observation and prisoners, about the lands west of the steppes. The Mongols would use that information fifteen years later.

By that time Genghis Khan was dead. So was Chepé, who had died on the return from his trip to Europe. But Subotai was still active. With Batu Khan, Genghis's grandson, he commanded a huge army that struck first at northern Russia, then moved south. This time, central and eastern Europe were aware of the danger and had mobilized. Boleslas the Chaste of Poland had assembled all his vassals. Henry the Pious of Silesia collected an army of Germans, Poles, Teutonic Knights, and a contingent from two other orders of military monks, the Knights Templars and the Knights Hospitallers. The king of Bohemia called up a force of Germans, Czechs, and Slovaks, while King Bela IV was waiting with 100,000 Hungarians. Subotai's scouts made him aware of every move of the Christian powers, while the kings of Europe knew only that the Mongols were somewhere in the vast and mysterious land of Russia.

In early spring, the Mongols moved west. Part of the Mongol army defeated King Boleslas and burned Kracow, first shooting the trumpeter who was trying to alarm the citizens. They destroyed the army of Henry the Pious before it could join the Bohemian army. Since the Bohemian army was too large for the Mongol divisions to attack, they tried unsuccessfully to lure King Wenceslas into a trap by ravaging the countryside. Finally, they tricked the king into marching north while they went south to rejoin the main body under Subotai and Batu. The main Mongol army then fell on the Hungarians. King Bela was thoroughly outmaneuvered and his army annihilated. The Mongols ravaged Hungary and burned Pest. Then they again left Europe and marched to the east. Ogotai Khan, Genghis's son and heir, had died, and a new khakhan—the supreme khan—had to be selected at a council of nobles. The Mongols had not been driven out of central Europe, they left of their own free will.

But they did not return.

The most formidable army of horse archers in history had been knocking on the gates of western Europe, but they never tried to complete the conquest. True, western Europe was not steppe, but that fact had never deterred the Cimmerians, Scythians, Alans, Huns, and all the other nomad peoples from the eastern grasslands who had pillaged their way across western Europe for centuries. It's also true that the empire of Genghis Khan was stricken by a horrendous outbreak of the Black Death, but that happened decades after Ogotai's successor had been chosen.[2]

Soil, Barbarians, and Feudalism

Although Genghis Khan proclaimed that "[t]here is one God in heaven and one Khan on earth,"[3] and his successors claimed to be emperors of all humanity, and although western Europe had been "inherited" by Tuli, khan of the Golden Horde, the Mongols never attempted to enforce their claim for several interrelated reasons.

One was the soil and rainfall. Western Europe, aside from semiarid areas of Spain, was ideal for growing crops, which meant a denser population. The farther east one moved, the less the rainfall and the lower the winter temperatures. Eastern Germany, Poland, and Russia had extensive swamps. Intensive agriculture was impossible there in the thirteenth century, so there was more open space and fewer people. The population of western Europe was much denser than that of Central Asia or eastern Europe.

But the population of western Europe was not denser than the population of northern China, and the Mongols and many previous nomad nations had repeatedly conquered northern China. There was an important difference between the populations of the two areas. Northern China was inhabited by a single people with a single culture. The waves of barbarians who had swept over it had been absorbed by the native population and become Chinese. Western Europe, however, had from prehistoric times been the home of warring tribes, many speaking different languages and following different customs. The Romans had at one time imposed a single government on much of the area. But the Romans never got to Ireland, the Highlands of Scotland, Scandinavia, the Baltic area, Russia, and much of Germany. When the Roman Empire fell, barbarians from the north and east overran the old boundaries and established kingdoms in what had been Roman territory.

Communications were primitive at this time and the organization of the barbarian kingdoms even more primitive. Most of the power in these new kingdoms was in the hands of local chieftains. Frankish kings were able to establish suzerainty over much of what is now modern France and Germany, but for most of the eighth century, it was a shadowy suzerainty. Charles Martel established a new dynasty and a strong Frankish kingdom, and Charlemagne extended Frankish dominion and created what was called the Holy Roman Empire. But this ghost of the old Roman Empire of the West broke up under attacks from all directions. Muslim pirates struck the Mediterranean coast and Moors crossed the Pyrenees to raid the south of France. Other raiders came from the east—Magyars, the horse archers from the steppes. Saxons and Danes attacked from the north. Worst of all were the Vikings, who raided the Atlantic, North Sea, and Baltic Sea coasts. Some of these pirates even sailed through the Straits of Gibraltar and struck the Mediterranean coast.

"Viking," incidentally, does not refer to a nationality or an ethnic group, although the vast majority of them were Scandinavian. It means "pirate" in Old Norse. There are references to Scottish and Irish vikings in the Icelandic sagas. An old English law decreed that anyone who "goes a-viking" shall be "nithing." In other words, his oath is no good, and every time he talks he is assumed to be lying. No one can do business with him, and anyone can kill him without paying a penalty. In spite of that, there were a number of English vikings, the last and most famous being Tostig Godwinsson, the brother of King Harold II, who was killed fighting for Harald Hardrada, the king of Norway, against his own brother during Hardrada's invasion of England. A few days later, King Harold II was also killed while fighting William the Conqueror.[4]

The impact of the Vikings was felt most strongly in Britain and Ireland and in northwestern France and western Germany. It greatly accelerated a continental social system that had been developing since the last days of the Roman Empire, the system we call feudalism. And that led to a new kind of fortification: the castle, which was probably the most important reason the Mongols had for avoiding western Europe.

The roots of feudalism go back to Roman plantations and Teutonic war bands. Toward the end of the Western Empire, taxes had become oppressive. To avoid them, some small landowners gave their farms to large plantation owners and became tenant farmers. The large landowners presumably were able to shoulder the burden of taxation better and may have had better political connections. The Teutons also had a protofeudalism. A prominent leader would gather warriors around him who would swear loyalty to him. The warriors were not necessarily from the chief's tribe, as can be seen in the Anglo Saxon saga of *Beowulf*.[5]

These raiding war bands were a further inducement for Roman peasants to put themselves under the protection of the big landowners. Most of these magnates had armed bodyguards, called *Bucellarians*, or biscuit eaters, which in some cases amounted to private armies.[6]

None of these developments—seeking the protection of the powerful in return for land, oaths of loyalty between the powerful and the less powerful, and the keeping of private armies—lessened during the chaos that followed the establishment of the Teutonic kingdoms. There were few cities in the West, and the population of those that did exist was well below what it was in Roman times. Most people lived in rural hamlets, and the king was far away. The local lord gained more and more authority. He also gained more and more wealth, which enabled him to afford such expensive items as horses and armor.[7]

Then the Vikings appeared. They took their shallow-draft boats far up rivers. Sometimes, they abandoned the boats for a time and stole horses so they could go raiding far from water. They stole every-thing that was movable, burned villages to the ground, and killed or enslaved the villagers. The villagers, who were poor subsistence farm-ers, had no way to resist these robbers.

The lords did, however. They built private fortresses, primarily, at first, with the forced labor of the peasants. The lords and their families lived in these castles. Some of their armed retainers lived inside the outer walls. And in emergencies, like a Viking attack, the peasants crowded in behind the walls. The castles, especially as they provided a secure base for the lord and his mounted and armored knights, proved an effective check on the Vikings. Castle building spread throughout western Europe. In France alone, there were 10,000 of these private forts.[8] There were many, many castles, because just about every chieftain who controlled a village or two built himself a castle. "It is this integrated combination of residence and fortress that is peculiar to the castle . . . and differentiates it from all other known forms of fortification, earlier, later or, indeed, contemporary," writes R. Allen Brown, an internationally acknowledged expert on castles.[9] The most similar forts are the castles of Japan, which were built long after castles had become militarily obsolete in Europe. But the Japan-ese castles were the private fortresses of only great lords, not of every samurai who could bully enough peasants to dig ditches and raise stone walls. Also, the Japanese lords usually lived in their castles only when an attack threatened.

Castles not only provided a defense against invaders, they also helped keep a conquered people in subjugation. When William the Conqueror defeated Harold, his knights spread all over England and forced the natives to build castles for them. At first, these were rather crude—a wooden house with thick walls and small windows set on a steep mound of dirt and surrounded with a palisade was typical. The mound was surrounded by a deep ditch and another palisade. And around that, or more usually, beside it, was a yard called a bailey inside another stockade. The bailey contained workshops and quarters for some of the lord's retainers and provided an emergency shelter for the peasants and animals. A drawbridge crossed the ditch and led from the bailey to a walkway up the mound to the lord's castle.

As soon as possible, the lords replaced these wooden forts with castles of stone. In some cases, they renovated old Roman fortifications. As time went on, though, castles became increasingly elaborate. The lords hired master masons—the same self-taught architects who constructed the great Gothic cathedrals—to build their castles. The master masons took many of their ideas from the Byzantine fortifications the crusaders had encountered and from the immensely strong castles the crusaders themselves built in the Holy Land.[10] By the time of the Mongol invasion, the majority of western Europe's castles were strong indeed.

Although the Mongol armies were all cavalry, and cavalry can't charge stone walls, the Mongols took many fortified cities in China and Central Asia. They used Chinese catapults and other siege engines, and they drove captives ahead of them when they stormed the walls. But with their excellent intelligence service, they knew all about those 10,000 castles in France and the thousands more in Germany, the Low Countries, and Spain. There were just too many.

The Europeans were at least as good as the Mongols in besieging and assaulting fortresses. They'd had several centuries to practice the art. And warfare in Europe was very slow going, even though no European potentate tried to take on *all* the castles in the area. The king of England did try to take on all the castles belonging to the king of France. The result was the Hundred Years' War, and the English king never came close to achieving his objective. The English strategy eventually degenerated into holding some castles and using them as bases for long-range surprise raids on French positions. The English won three great field battles resoundingly, but the French won the war by retaking, one by one, all the English bases. It was not safe to bypass a hostile castle, because each castle was a base for mounted knights who could play havoc with an enemy's communications and supplies. The

Mongol army tried to live off the land, but small foraging parties would be extremely vulnerable to the garrisons of hostile castles.

The castles provided one major turning point in war: they set a limit on the hitherto invincible Mongols and ensured that west European culture would survive. They also created a second turning point, one that would make possible the military supremacy of the Western nations. The many castles with many diverse masters ensured military competition, not only among the lords, but also among the lords and a new, increasingly powerful class: the merchants.

5

Private Armies and Public Policy

Trained mercenaries like this crossbowman were the first professional soldiers and the forerunners of regular armies.

Warriors and Businessmen

THE DARK AGES were especially dark for merchants. Moving merchandise anywhere but between coastal cities involved either caravans or river boats. It also involved paying a tax to whoever controlled the land the merchandise was crossing. In Asia or Africa, this was a major annoyance, but the merchants accepted it as a cost of doing business. European merchants weren't so accepting. And when they shipped products by sea, the merchants had to be prepared to cope with pirates. Many merchants in European coastal cities were actually reformed pirates, and some were still practicing. After all, pirates had to find a way to exchange what they stole for something they found more desirable.[1]

William H. McNeill, in *The Pursuit of Power*, suggests that north Europeans, including merchants, were more violent than their counterparts in China and India, because they were more inured to bloodshed. In Europe, especially in the north, it was the custom to slaughter most of the pigs, sheep, and cows in the fall and cure the meat. Otherwise, the livestock would eat up all the stored fodder long before spring. In northern China, livestock was not so plentiful, and many Indians would consider such slaughter a sacrilege.[2] The idea that routinely shedding blood desensitizes people to killing is also presented by Lieutenant Colonel Dave Grossman, a former U.S. Army psychologist, in his descriptions of modern training in the U.S. military.[3]

As a result of this indifference to violence, McNeill contends, European merchants did not tamely pay the various tolls demanded by the lords. Instead, they armed themselves and recruited armed retainers. However, there may have been more to the merchants arming themselves than the violence-prone nature of European society. There were so many lords in so many castles that if the merchants had given each what he demanded, every caravan would result in a loss rather than a profit.

So in early feudal times there were two groups who were armed and ready to fight: the knights and the merchants. At first, the knights were stronger. But since the conduct of commerce in Europe was not a continuous battle, the two armed groups must have worked out some arrangement. The merchants did not get free travel, but the knights didn't get nearly what they thought they should have. As time went on, the balance of power slowly shifted toward the merchants.

This shift came about because the commercial interests began to concentrate in certain strategic towns. Italy was the first home of such towns. Northern Italian seaports were ideally placed for sea trade with the Byzantine Empire, Egypt, the Levant, and North Africa and also for sending caravans across the Alps. Far to the northwest, similar commercial centers developed in the Low Countries, where the Meuse, Scheldt, and Rhine Rivers converge.[4] These towns depended on trade, not agriculture, and eventually they would brook no interference from the castle lords. The citizens built walls around their cities and formed militias of citizens to defend them. A crowd of crossbow-armed burghers atop city walls would cause any knight to have second thoughts.[5] In Italy, where the most advanced cities were located, the citizens began to take the field with pikes and cavalry as well as crossbows.

In 1176, 110 years after William the Conquerer's combination of cavalry, infantry spearmen, and infantry archers defeated the infantry of King Harold Godwinsson, infantry pikemen and crossbowmen at Legnano trounced a force of German knights who were trying to enforce the rule of Frederick Barbarossa, the Holy Roman Emperor, over the cities of Italy. The victorious infantry were militia of the Italian cities in the Lombard League.

Incidentally, in spite of loose chatter about a "cavalry cycle" in the Middle Ages, infantry played key roles all through the period. From Hastings in 1066, through the Crusades, Legnano, Courtrai in 1302, Crécy in 1346, Poitiers in 1356, to Agincourt in 1415, infantry defeated cavalry in the field. Scottish infantry spearmen under William Wallace routed English knights at Cambuskenneth Bridge in 1297. Wallace and his spearmen were later beaten at Falkirk by English infantry archers, assisted by English knights. The Scottish infantry under Robert the Bruce again triumphed at Bannockburn in 1314. The English chivalry suffered their greatest ever loss of life in a single day, and Bannockburn ended English attempts to conquer Scotland. Infantry, of course, were the main forces in all of the many sieges of this period, both as defenders and attackers. So it is simply not true to say that the Swiss pikemen at Morgarten (1315), Sempach (1336), and Näfels (1380), or the English archers at Agincourt ended the cavalry cycle. The cavalry cycle never really existed. What did exist—and always has—was the superiority of trained troops, whether on foot or on horseback, over untrained multitudes. All knights had studied fighting from the time they were children.[6]

And when city militia were able to rout knights, war had become much more complicated. It required the cooperation of three types of soldier: pikemen, crossbow archers, and cavalrymen. Pikemen provided a mobile wall that protected the crossbow archers and the city cavalry until they were ready and able to engage. The crossbows provided the firepower to knock enemy knights off their horses. And the city cavalry would come out in the open to drive the demoralized foe from the field. A break in the line of pikemen or panic among the crossbowmen could mean disaster. The militia had to be well trained.

This posed a problem for the city militias. Artisans and merchants could not afford to spend all their time practicing for war. And as the commercial enterprises got bigger and trade became more widespread, divisions developed. Purveyors of spices and armor, for example, might have conflicting interests. Employees and employers usually did. And as cities like Venice and Genoa began to assert control over the hinterlands, they needed troops to guard frontiers. To meet their new needs, the cities turned to mercenaries.

The Condottiere

Mercenaries had long been a fixture in Europe. Feudal commitments bound a vassal to serve his overlord for only a limited time, even if the overlord was the king. After that, the king had to pay his troops. Mercenary soldiers who were always fighting looked like a better buy than knights who were mainly occupied in supervising their estates—and a much better buy than peasants who were recruited from the fields. And the knights, who were increasingly involved in local government and other enterprises, were often happy to give the king money in lieu of military service. The English kings were major employers of mercenaries. They hired Welsh archers for their campaigns in Scotland and Welsh and English archers for the battles of the Hundred Years' War. The seemingly interminable wars in Italy between Spain and France also involved thousands of mercenaries, some of them North African Muslims.

In one of those Italian wars, Pedro the Great of Aragon defeated Charles, the French king of the Two Sicilies. When Charles and Pedro signed the Peace of Caltabellotta in 1302, most of the Aragonese army became unemployed. One of them, Roger de Flor, organized these jobless warriors into what was called the Grand Company or the Catalan Company. It was a mixture of heavy and light cavalry, and infantry pikemen, sword-and-shield men, and crossbow archers, with a large proportion of the last. It was actually an army without a country, 5,500 men in all. The Catalan Company engaged in the wars in the Balkans for various lords and consistently defeated French, Greeks, Turks, Spaniards, and Germans for the next eighty-seven years.[7]

The Hundred Years' War became a great source of mercenaries. The English and French kings periodically made truces, which left their soldiers unemployed. Some of these troops merely stayed in France and became bandits, plundering the peasants. Others crossed the Alps into Italy, where they entered the service of the Holy Roman emperor, the king of France, or one of the Spanish kings. When they were without contracts, they plundered the countryside. It was to these hardened warriors that the Italian city-states turned.

The city fathers signed contracts—*condotta* in Italian—with the captains of mercenary companies, who were called *condottiere*. One of the best known was a veteran of the Hundred Years' War called Giovanni Arcuto. This was as close as his employers could come to his unpronounceable English name: John Hawkwood.

At first, the cities hired mercenary companies for specific campaigns. As the "free companions" saw it, today's employer might become tomorrow's enemy. Today's enemy troops might become tomorrow's allies. Consequently, the free companies tried not to spill too much of each other's blood. War became a rather profitable game for the mercenaries, but the cities often felt they were not getting what they paid for. Besides, there were so many wars that short-term contracts proved to be uneconomical. The cities began offering long-term contracts. A captain would agree to raise troops who would serve for a period of years whether there was a war or not. The captain would receive regular payments, and he would pay the troops. He and his troops would also be allowed to keep some of the plunder of enemy lands. This job security looked good to the mercenaries, so contracts were usually renewed again and again. Lifetime service became common.

If the situation was good for the mercenaries, it was even better for the citizens. Most of them agreed that paying a regular tax to hire soldiers was preferable to being plundered. But there were greater benefits. When they were plundering, the mercenaries took cattle and sheep, wheat and fruit, clothing and firewood, as well as gold and jewelry. Most of the plunder was consumed and passed out of existence, and the gold and jewelry were usually sold far from where they were stolen. However, when the citizens hired the mercenaries, they paid them money. The soldiers took the money and bought what they needed from the citizens. Unlike plunder, payments to the mercenaries increased the city's business.[8] There is no doubt that the canny merchants and bankers of northern Italy considered this aspect of the situation.

The trouble with condottiere was that sometimes one of them would decide that taking the city's pay wasn't enough, so he'd take the city. That's what the condottiere Francesco Sforza did in Milan. The Republic of Venice avoided this problem by hiring mercenaries in small units. A free company at that time was divided into *lances*, units similar to a modern squad—three to six men who formed a combat team. They were usually armed differently and trained to support each other. The Venetians made contracts with individual lances. The city was then able to combine the lances into larger units and appoint officers to command them. The prospect of promotion also inspired the junior officers to excel at soldiering.[9] It was a short step from the Venetian military system to a standing army—something western Europe had not seen since the Roman Empire.

Better Weapons

In most areas, such as China, the government controlled weapons production. Even where it didn't, as in Japan, warriors were happy with only traditional weapons. Oda Nobunaga rose to power using guns, but under the Tokugawa shogunate, the manufacture of guns was gradually eliminated.[10]

This was not the case in western Europe. The area was full of city-states, duchies, and principalities competing militarily, with semi-independent military units serving them. Each wanted weapons that would give it an advantage over its competitors. And the independence the commercial classes of western Europe had rather painfully established ensured that there would be many suppliers competing to give them those weapons. European metalworkers learned to make wrought iron and steel in large enough quantities to manufacture magnificent plate armor. Plate armor was a necessity in western Europe because arrows from longbows and crossbows could penetrate mail almost as if it wasn't there. Arrows from the English longbows could pierce even plate armor at short range, and bolts from the most powerful crossbows could penetrate it at far longer range. Crossbow makers developed infantry crossbows with a pull weight of more than a thousand pounds, which could be bent only by using an arrangement of pulleys and a windlass.[11]

This European search for the ultimate weapon was dramatically demonstrated by what the Europeans did with an invention that had been brought to the West by the Mongol conquest. It originated in China, where it was primarily used for entertainment. In the West, it brought about one of the greatest of all turning points of war.

6

The Devil's Snuff: The Gunpowder Revolution

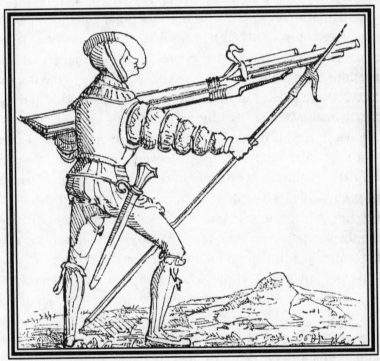

Soldier with a hand cannon. Weapons like this were an early application for gunpowder.

Stillborn Revolution

IT WAS 1575, and the future of Japan lay in the hands of two men: Takeda Katsuyori and Oda Nobunaga. The Takeda family was among the most powerful of the *daimyos*, or great lords, of Japan. They owned the prime horse-raising land in the country and were famous for their cavalry. They were sometimes referred to as "the invincible Takedas."

Oda was born into relatively minor nobility, but he was becoming the most powerful of the daimyos in the chaotic and interminable civil war the Japanese called the Century of War. Oda had just deposed the shogun, who appealed to the Takedas for aid. Both Oda and Takeda Shingen, the father of the daimyo who was now besieging Oda's Nagashino Castle, had something in common. They had once both depreciated the value of a weapon that had been introduced a generation before: the gun. And both had changed their minds.

Takeda Katsuyori agreed with his father that guns had their uses, but he placed most of his faith in his magnificent cavalry. Oda, however, saw guns as the wave of the future. Or at least, that's what several generations of Japanese writers passed on. According to the story, Oda had 10,000 musketeers in his army, with 3,000 specially trained in tactics he had invented.[1] Noel Perrin, in *Giving Up the Gun*, tells the traditional story. Oda divided his 3,000 shock troops into three parties. He put the first thousand behind a stockade, the second in a trench, and the third behind a breastwork in the town of Shitaragahara, near Nagashino Castle. When Takeda's horsemen charged, the first group was to fire, run to the rear, and reload so it could fire again. Meanwhile, the second group would fire. It, too, would run to the rear to reload while the third group fired a volley.

Takeda's "invincible" cavalry charged, and in a rapid succession of volleys, it was mowed down like ripe rice, as the story goes. Perrin quotes a Japanese lieutenant general who wrote in 1913 that little improvement in infantry tactics had been made since Oda's victory at Shitaragahara. As it demonstrated in World War II, the Japanese infantry certainly must have learned a lot since 1913. Oda's tactics at Shitaragahara closely resemble the British street-fighting formation used—with a notable lack of success—at Concord in 1775.[2] But he may not even have used those tactics. Mitsuo Kure, a modern Japanese historian, says research indicates that Oda had only about 1,000 gunners in his army, which, combined with that of his vassal Tokugawa Ieyasu, numbered 30,000. Meanwhile, Takeda had no more than 15,000 soldiers,[3] and he attacked an entrenched army twice his size. If he hadn't been defeated it would have been a miracle. Also, Takeda's horsemen dismounted to fight. The samurai had learned that a man on horseback made a better target for the rather inaccurate guns of the period, so for the last few years, the mounted bowman and lancer had morphed into the infantry gunner and spearman.[4] But for all that, Takeda was decisively defeated; Oda became the ruler of all Japan, and guns played a major role in his victory.

Oda's rise was a decisive event in the history of Japan. There are some other aspects of the Battle of Shitaragahara (or the Battle of Nagashino, as it's sometimes called) that are more interesting in world history.

First is the remarkably long time it took for gunpowder to be used in Japan. Gunpowder had been in use in China since about A.D. 1000.[5] At that time, it was used for firecrackers and other pyrotechnics, but 100 years later, the Chinese were throwing hand grenades at their enemies. These were paper or bamboo tubes filled with gunpowder and stones, broken porcelain, or iron bullets. They were followed, around 1290, by bamboo guns—tubes with one end open from which a charge of gunpowder fired the stones, bullets, and so on at enemies.[6] That was a long time before Takeda and Oda had their shoot-out. What makes the delay even stranger is that gunpowder was carried to Japan almost three centuries before Japanese samurai began shooting at each other. The Chinese-Korean-Mongol fleet of Kublai Khan that attempted to invade Japan in 1274 and again in 1281 was equipped with gunpowder bombs the Mongols intended to shoot from mechanical artillery. In the first invasion, the Mongol army of 40,000 easily drove back the 120,000 samurai who opposed it. The Mongols' discipline and archery were factors, but so were the gunpowder bombs and rockets they shot at the Japanese. Fortunately for the Japanese, the Mongol fleet was destroyed in a typhoon. In 1281, Kublai Khan tried again. This time he sent 150,000 men who were again equipped with bombs and rockets. The Japanese resistance was stronger this time. While the Mongols were fighting on land, small Japanese boats attacked the Mongol ships and set fire to some of them. The Mongol commander decided to move the fleet. But as the ships were getting underway, another extraordinarily powerful typhoon struck. The ships were dashed into each other and into rocks on the shore. Almost all of the invasion fleet was destroyed.[7] The Japanese, understandably, saw those two typhoons as intervention by the gods—kamikaze (divine wind). Recent exploration of the invasion ships wrecked in the last kamikaze typhoon definitely establishes that they were equipped with ceramic gunpowder bombs.

Once they adopted guns, the Japanese went all the way. In a way. "At least in absolute numbers," Perrin writes, "guns were almost certainly more common in Japan in the late sixteenth century than in any other country in the world."[8] Oda's successor, his henchman Toyotomi Hideyoshi, invaded Korea in 1592 with 160,000 men. More than 40,000 of them carried matchlock guns.[9] Toyotomi, who had no lack of ambition, had planned to conquer Korea first, then China, and then the Philippines.

His campaign hit a snag at the beginning. Yi Sun Shin, a Korean admiral, had his ships tow several turtle-shaped vessels into Pusan Harbor, where the Japanese fleet was anchored. The "turtle ships" had shells of iron and rows of cannons along their sides. Yi himself had designed the turtle ships. They sank sixty Japanese ships. In spite of popular belief, the duel between the ironclads *Monitor* and *Virginia* (nee *Merrimack*) was not the first use of ironclad ships in combat, as the Koreans introduced ironclad warships 270 years earlier. The Japanese, in spite of this setback, pushed the Koreans up the peninsula. Then China intervened. The Chinese had invented gunpowder half a millennium earlier, but they had fewer guns than the Japanese. The Chinese, though, had more men. And then Yi and his turtle ships struck again. In November 1598, at the Battle of Chinhae Bay, Yi's ironclads sank 200 of the 400 Japanese ships, and the rest fled back to Japan. Yi, a greatly unappreciated naval hero, was killed during this battle, but he had saved his country. In his headquarters in Japan, Toyotomi, one of the three unifiers of Japan (the other two were Oda and Togugawa Ieyasu), died about the same time. The Japanese withdrew from Korea.

Toyotomi is one of the few Horatio Alger figures in Japanese history. He was born a peasant and became regent of Japan. But once he reached the top of the Japanese social order, he took measures to

ensure that no other peasant would follow in his footsteps. He con-
fiscated all the guns and swords owned by any who were not samurai.
Under Toyotomi and his successors, the Tokogawa shoguns, strife-
ridden Japan was to enjoy three centuries of domestic tranquility. But
there was a price. Except for the Inca empire, Japan became the most
rigidly controlled society in the history of the world.

The Tokugawa shoguns banned foreigners and virtually ended all
foreign trade. Now, they believed, there was no necessity for guns, so
the possession and production of firearms was increasingly restricted.
The samurai, who gloried in their skill with the sword and the bow,
detested the idea that some unskilled rural clodhopper with a gun
might kill one of them. When Commodore Matthew Perry came to
Japan in 1853, the few matchlocks still being made had not changed
from the design used by the followers of Oda.

The Fertile Soil of Western Europe

The Chinese had invented gunpowder, but they didn't see its military
potential. As noted earlier, they used it to make a rather weak sort of
hand grenade and an even weaker type of gun. They also used it for
rockets. The Mongols shot ceramic bombs from their siege engines. All
of these weapons, however, were mostly useful for frightening horses.
The gunpowder mixture was weak, as were the walls of the bombs
and the guns, so that even if a more powerful explosive were used, it
could not be contained long enough to build up much pressure. Lieutenant Colonel H. W. Hime, an expert on the subject, concludes that
the first gunpowder mixtures contained unrefined potassium nitrate,[10]
which would have prevented them from developing enough pressure
to propel bullets with much velocity.

Confucian scholars, from the emperor to the lowest mandarin,
ruled China, and Confucius had little use for either soldiers or busi-
nessmen. There was no desire to create better gunpowder even if
someone had thought it was possible. The Mongols who conquered
China did not, of course, hold soldiers in contempt, and they were
open to new ideas. They adopted the Chinese *huo-pa'o*, a gravity-
powered siege engine the Europeans called a trebuchet, to shoot

gunpowder-filled bombs into besieged cities. But to the Mongols, gunpowder was mostly for sieges. Their military strength was built on mounted archers, organized in decimal units that were highly disciplined and responded to signals. They could see that gunpowder could frighten horses and confuse enemy soldiers, but it wasn't a major weapon.

The Mongol conquest of much of Eurasia, however, almost certainly led to the supremacy of gunpowder weapons. The Mongols used gunpowder in their western campaigns, and after the conquest, trade and travel in the vast empire ruled by the great khans increased enormously. Other than in Chinese records, there is no mention of anything that could be construed as gunpowder before the thirteenth century. Medieval Arab documents call potassium nitrate, the essential ingredient of gunpowder and other pyrotechnic mixtures, "the snow from China." Regardless, the Arabs, like the Mongols and Turks, saw little military use for gunpowder. All three peoples considered the horse archer the ultimate weapon.

The earliest European reference to gunpowder is in the writings of the monk Marcus Graecus. Most of the manuscript is in Greek and apparently written in 846. But part of it is in Latin, a type of Latin not in use before the mid-thirteenth century. The Latin section gives thirty-five recipes for gunpowder—mixtures useful for everything from rocket propulsion to blasting. A letter from the same period by a Spanish monk named Ferrarius mentions gunpowder, as does a manuscript by the medieval bishop, philosopher, and scientist Albertus Magnus. The most definitely dated writing is a letter from Friar Roger Bacon to the bishop of Paris. In it, Bacon not only gives a formula for gunpowder, but also directions for the refining of potassium nitrate, also called saltpeter. This is the first known description of the refining process.[11] Bacon wrote his letter in 1252. That was just thirty years after Subotai Bahadur and Chepé Noyon invaded Georgia, and just eleven years after the Mongols burned Krakow and wiped out the army of Bela IV.

Western Europe at this time was more receptive to new ideas in warfare than anywhere else on the globe. The Crusades were turning into a disaster: European chivalry was just unable to cope with the

Turkish and Arab horse archers in the desert. In Europe, mercenary captains were looking for anything that would give them an edge on their enemies. The many competitors in the armaments industry were looking for new and better products for the captains and princes who were their customers. Even the alchemists were engaging in more practical pursuits than trying to find the process for turning lead into gold. Someone—almost certainly a European—had discovered the need for refining potassium nitrate, and Bacon had reported on the process.

The more powerful gunpowder now available could be used for demolitions. Its first use was in siegecraft. In the past, attackers undermining a wall would prop the bottom of the wall up with timber supports as they dug. When the mine was completed, they'd set fire to the supports and wait for the wall to collapse. Now, they just put gunpowder under the wall and blew it up. Another gunpowder device was the petard. This was a heavy iron pot filled with gunpowder and hung on a castle gate. The "engineer" ignited a fuse on the petard and ran for his life. The weight of the pot tamped the explosive. When the charge went off, it blasted the door in and sent the petard flying in the other direction. If the engineer wasn't fast enough, he would be "hoist with his own petard."

The petard may have suggested the gun. If the pot were somehow made stationary, a missile placed over the powder would fly out. A device like that had possibilities, and using it would be much less dangerous than trying to nail a petard to a gate, then getting back to your own lines without being hit by enemy arrows or your own petard. Experiments along these lines showed that the missile would strike much harder if the pot were elongated.

Some of the experimental guns looked quite odd. The earliest picture of a gun is on a manuscript by Walter de Milemete, tutor to the boy who would become England's King Edward III. It shows a vase-shaped object, lying on its side on a four-legged table while a man in armor is apparently igniting the powder in a touch hole. Projecting from the mouth of the vase is a short, heavy arrow, complete with vanes. De Milemete's picture dates from 1326. Another picture of a similar gun with four armored soldiers serving it is also ascribed to de Milemete, although it may be even older.[12] And there's a picture of a very similar gun, dating from 1332. But this one comes from China.[13] Apparently, new ideas traveled in both directions across the great khan's empire. Although the gun looks like a vase, it probably had a cylindrical bore. A similarly shaped gun, dating from the mid-1400s and found in Sweden, has this type of bore.[14]

At first, small cannons were made for fieldwork. The English, led by Edward III, were reported to have used two or three in 1346 at the Battle of Crécy in the Hundred Years' War, but their performance was greatly overshadowed by that of the English archers.[15] A contemporary account described the English guns as weapons "[w]hich with fire throw little balls to frighten and destroy horses."[16] Even though the guns weren't that accurate, some military historians suggest that the blasts from the guns may have demoralized the mercenary Genoese crossbowmen in the French army.[17] Even earlier, the records of Florence show that firearms were in general use there in 1324.[18] Many of these arms were small and could easily be cast in bronze. That's also true of the many barrels of the "ribaudeiquin" used by Edward III in 1346. This gun was capable of firing a volley of bullets, but it took a long time to reload all of those barrels.

Small guns in those days were not nearly as effective as the longbows and crossbows that armies had been using for centuries, because they didn't have the range, accuracy, or penetration and they were slower to load. Sieges were another story. Bombards, huge guns throwing stone balls weighing hundreds of pounds, were more effective than catapults or ballistae, engines that use the elasticity of twisted rope to throw missiles. They even outclassed the trebuchet, a machine mounting an enormous beam with a weight of many tons on the short end and a sling on the long end.

There were two ways of making a bombard. It could be cast in bronze or be built of wrought iron. The second method involved welding iron rods together around a mandrel, then welding iron hoops around the rods. The resulting tube was open at both ends, but the breech was closed by wedging an iron powder chamber to that end. Both types of bombard were expensive—the bronze, because so much of that expensive metal was used; the iron, because of the labor involved. Only kings and great lords could afford bombards. Their introduction was a big step toward the centralization of European countries. Europeans had a virtual monopoly on the production of huge bronze bombards, which, being safer than the built-up iron guns, became more common. The European foundries had learned the difficult art of making gigantic castings over decades of casting great bells for their cathedrals.

When Mohammed the Conqueror attacked Constantinople in 1453, he hired a renegade Hungarian named Urban to cast bombards bigger than anyone had ever seen before. The guns would be so heavy that Urban told the sultan it would be easier to cast them on the spot than move them from a foundry. Mohammed consigned his small guns to the founder's furnace, and Urban began casting. A similar gun on display at the Tower of London is seventeen feet long, has a bore of twenty-six inches, and weighs almost nineteen tons. It fired a stone ball weighing 800 pounds. Some of Mohammed's bombards were reported to fire stone balls weighing as much as 1,500 pounds. Like all cannons of that period, they fired from a fixed wooden mounting that was built around the gun. Incidentally, Mohammed did take Constantinople, but the guns didn't do the job. They battered down sections of the city's outer wall, but each time the Turks attempted to enter the breach, they found that the Greeks had built another wall behind it. The city fell after some Turkish janissaries found an undefended postern gate, entered it, and attacked the defenders from the rear.[19]

Castles and Guns

Castles of the fourteenth century had high stone walls to make it difficult to get over them with ladders or siege towers. They were often surrounded by a water-filled moat, which made it impossible, given the engineering capabilities of the time, to dig under the walls. Castles like that could only be destroyed by cannons. After a few mighty fortifications had been blasted to rubble, owners of forts decided that something had to be done.

To reduce the size of the target, some castle designers lowered the walls and lowered the towers even more, so that they were level with the walls or only a little higher. Sometimes, they filled their towers with earth so their own cannons would have a firm base from which to fire. Few towers, however, were wide enough to support a fair-sized cannon and few could withstand the continued shock of the gun's recoil. Lowering the towers and walls did reduce weak spots, though. Because the thickness of the curtain walls and sides of the towers tapered, the thin masonry at the top of fortresses was especially vulnerable to cannon fire. Actually, all masonry was vulnerable. To reinforce the stone work, garrisons banked up earth behind the walls. The trouble with that was that the earth exerted pressure on the wall from behind. When a wall collapsed under bombardment, the earth poured through the gap, giving the besiegers a convenient ramp to get through the breach.

Things got much worse for defenders of castles after the French and Burgundians developed their new mobile siege artillery. During the Hundred Years' War, the French and the Burgundians, English allies, had an artillery arms race. The English, apparently convinced that their longbow was the ultimate weapon, did not enter the race. The race resulted in a new type of siege gun. It was of much smaller caliber than the big bombards and it fired iron, instead of stone, cannon balls. Iron is denser than stone and much tougher, so an iron ball fired at the same velocity as a stone one does far more damage: the stone balls usually shattered on hitting masonry, while iron cannon balls were often intact enough to be reused. And the new guns fired their shot at a much higher velocity, because the guns were proportioned so that they could handle higher pressures than the bombards. Their barrels were longer in proportion to the bore, so the expanding gases from the explosion had more time to act on the shot, greatly increasing the velocity. Most important, the new guns had wheels. They could be towed into position and fired immediately. They were also cast with a pair of lugs, called *trunnions*, just in front of the center of gravity. The trunnions made it possible to pivot the gun in its carriage. The gun muzzles could be elevated or lowered, and the guns could be traversed by moving the long "trail" of the gun carriages. While they were improving guns, the artillerymen were also improving gunpowder. Early gunpowder was simply ground fine and mixed. The three components, saltpeter, sulfur, and charcoal, tended to separate when the powder had been stored for a time, which resulted in a less powerful explosion or no explosion at all. Now, the components were mixed while wet, then the damp powder was separated into "corns," each of which contained the three components in the correct proportion, and was permanently bonded.[20]

With this new artillery, the French in 3 years drove the English out of all the castles and fortified towns they had acquired over 113 years of warfare. Then King Charles VIII took his guns into Italy to enforce his claim to Naples. He fought his way down the peninsula and back, terrifying all the Italian cities in his way. The Italians took a new look at the art of fortification. They had plenty of incentive. France's incursion into Italy brought Spain and the Holy Roman Empire (mainly Austria) into the peninsula to counter it. The Italian city-states and the Papal States joined the fray. The Italian Wars provided plenty of opportunity to experiment with systems of fortification.

In the early part of the wars, the attackers went from victory to victory. In 1500, the French got a surprise when they besieged the little city of Pisa. The Pisans had thrown off the yoke of Florence, a French ally, and the French were going to teach them a lesson. They hauled their formidable siege guns down to Pisa and promptly battered down part of the city wall. But when they rushed through the breach, they found that the Pisans had built another barrier behind it. The Pisan *retirata* consisted of a deep ditch with perpendicular walls in front of a thick earth rampart. Behind the rampart, Pisan artillery mowed the attackers down. The rampart was a small target for the French gunners, and when they hit it, their cannon balls seemed to do no damage. Earth walls didn't shatter like stone walls: they just absorbed the shot. Nine years later, Venice defied not only the French, but the Austrians and Spanish as well. Three of the four continental superpowers attacked Padua, Venice's foothold on the mainland. (The fourth power, Turkey, wasn't involved, although the Turks and Venetians were ancient enemies.) But Padua's wall was, like Pisa's, backed by a ditch and earth rampart. The rampart was more elaborate than Pisa's hastily constructed fortification and was reinforced with timber. This fortress frustrated the most powerful armies in Europe.

Earthworks—ditches and mud walls—sprang up all over Italy. Some of them were designed by the most famous engineers of the day, people like Michelangelo. To protect Florence, the great sculptor and painter diverted a stream and built earthen bastions to guard the city gates. More important, he built a string of detached earth-and-wood forts on the surrounding hills. The bastion was a replacement for the tower. It was a building shaped like an arrowhead with the point facing the likely position of the enemy. It was low, wide, and surrounded by a deep, wide ditch, and it mounted cannons. Built along city walls, bastions provided flanking fire. Detached fortifications like Michelangelo's could disrupt an enemy attack before it reached the city walls.

Earthworks proved to be formidable fortresses, but princes and cities yearned for permanent fortifications. Earthworks needed constant attention so they wouldn't eventually be washed away in the rain or destroyed by vegetation. So the engineers translated their developments into stone—or more precisely, stone and earth. The ditches were faced with stone. The curtain walls were of both stone and earth, but with internal stone buttresses to help the facing wall resist the pressure of the earth behind it. The bastions were constructed the same way. The way the "facings" of the bastions were angled eliminated the "dead ground" that could not be covered by fire from the wall or other bastions. Dead ground was a major weakness of the round tower. More cannons could also be fired from the straight sides of a bastion than from a round tower of similar size. Engineers soon changed the "flanks" of the bastion, which connected the facings and the curtain wall. They "retired" the flanks by pulling them back behind the facings so they could not be seen from the front. The guns in these flanks, sometimes on two levels, could only be seen by someone in the surrounding ditch, or "dry moat." This way, there was little chance of an effective counter to them: moving an enemy cannon into the dry moat would be most difficult.

To further protect the fort, engineers built a bare, sloping earth embankment in front of the dry moat. It almost hid the walls and bastions and provided no cover for attacking infantry. Just behind this *glacis* and on the forward edge of the dry moat was a path for infantry called a "covered way." Troops on the covered way could fire on attackers and, on certain wide spots, assemble for counterattacks. To further strengthen the fortress, engineers built detached fortifications, some inside the dry moat, some with moats of their own. They were connected to the main fortress by drawbridges or tunnels. If attackers overwhelmed them, but missed the tunnel entrance, they could be reoccupied to fire on the attackers from the rear. The Italian engineers countered the threat of mines with a number of methods of locating enemy excavations. Sometimes, they buried drums half into the ground and placed dried peas on the drumheads. The vibrations from digging made the peas bounce on the drumhead. The peas that bounced most were those nearest the mine. Or they attached rattles to the wall. When they determined the direction of the mine, they began a countermine. Counterminers attacked the enemy with firearms or with burning oil, or they mined the enemy mine and blew it up. Sometimes, they dug countermines in front of their fortresses at logical points of attack before any enemy approached.[21]

Most of the development of this new type of fortification was done in Italy, and Italian engineers were in demand all over Europe. Italians, then other Europeans, continued to develop the bastioned fortress, which was the standard military strongpoint until well into the nineteenth century.

Guns on the Battlefield

While guns were revolutionizing siegecraft, they were also making tremendous changes on the battlefield. In open fighting, guns, as we have seen, were initially hardly more effective than the first Chinese guns. But the Chinese had no interest in developing guns, while the Europeans did. The first infantry guns—"handguns," in the terminology of the time—were miniature bronze cannons mounted on the end of a long pole. The weapon looked something like a spear with a short tube where the spearhead would be. But in a very short time, gunsmiths recognized the value of a much longer barrel. As the barrel grew longer, the wooden mount got shorter. Finally, the wooden shaft came to resemble a modern gunstock. The gunner fired his weapon by touching a piece of smoldering cord, called a "match," to a tiny hole in the rear of the barrel called a "touch hole." This was rather awkward, so gunsmiths added a metal matchholder to the gunstock that could be lowered to the touch hole with a trigger. These early matchlocks were not as accurate as a crossbow and were slower to load than a longbow, but they had one great advantage: they could pierce heavier armor than the older weapons.

Before gunpowder became a major presence on the battlefield, the most copied infantry formation was the Swiss phalanx. This was a dense mass of infantry, lightly armored and armed with pikes and halberds. The pikemen were in the front ranks to stop any cavalry charges. When they did that, the halberdiers, using a six-foot-long combination of ax, spear, and pick, moved to the front and chopped up the stalled enemy cavalry. To provide missile support, crossbow-men marched beside the phalnax. When matchlocks were developed, the Swiss and their imitators replaced the crossbows with guns. But other than that, the phalnax remained unchanged. It would, of course, have made a perfect target for artillery, but most artillery in the early sixteenth century was not mobile enough to use on a battle-field. Once in a while, though, the big guns saw action in an open battle. In 1522, Prospero Colonna, the commander of the imperial forces, was besieging Milan, a French dependency. The French tried to raise the siege, but Colonna had built fortifications facing outward around the city, called "lines of circumvallation." The French com-mander hesitated when he saw the imperial position near the town of Bicocca, a breastwork behind a sunken road. But his Swiss mercenar-ies, who considered themselves invincible, began to grumble. Fearing a mutiny, Lautrec, the French general, launched his Swiss against the besiegers on April 27. The result was horrendous. Colonna's cannon balls blasted bloody lanes through the massed Swiss. A single shot could kill as many as thirty or forty men. A thousand Swiss were killed before they reached the ditch. When they leaped into it, four lines of Spanish handgunners, each line firing a volley, shot them down. A few Swiss climbed over the bodies of their comrades and reached the top of the rampart, but the allied pikemen pushed them back. More than 3,000 of the attackers were killed.[22] The surviving Swiss fled, and as Christopher Duffy puts it, "The bellicose and independent spirit of the Swiss was broken forever."[23] The French were driven out of Italy.

Three years later, handgunners demonstrated that they could inflict a decisive defeat without the aid of cannons. King François I returned to Italy in 1524 with a large army and retook Milan and besieged Pavia. An imperial army, chiefly Spanish, arrived to break the siege. It was commanded by Fernando de Avalos, the marquis of Pescara, and the former constable of France, Duke Charles de Bourbon, who had joined the imperialists. The constable of Bourbon, as he was called, was the senior officer, but Avalos, a veteran of Bicocca, was the key man.

The imperialists charged the French on February 21, 1525, but the French artillery blasted them back. The Spaniards took cover. The French cavalry charged them, but the Spanish pikemen rose and held their ground. Then the handgunners Pescara had placed on the flanks opened fire. They were using a heavy, long-barreled matchlock that could knock down the most heavily armored man or a horse with a single shot. They cut down the French cavalry and the Swiss infantry. The Spanish charged again and captured all the French artillery as well as François himself.

The wars of religion were beginning at this time. In one of the first, the followers of John Hus, a Czech executed as a heretic in 1415, adopted guns in their war against the Holy Roman Emperor. They mounted their primitive cannons on wagons, supplemented them with very primitive handguns and crossbows, and rampaged through central Europe. When they confronted an enemy, they literally circled their wagons and blasted the attackers with missiles. As most of the imperial forces were cavalry, the *wagenburg* provided an excellent defense. When the enemy troops were sufficiently weakened, the Hussite cavalry emerged from the wagenburg and finished them off. The Hussites were undefeated for four years, until they tried to invade

Hungary and were repulsed. After that, they split into warring factions. They reunited and resumed their ravaging of Germany, Hungary, and Silesia, but they again split into warring factions. Worn down by civil war, they ended their rebellion.

Gunpowder and the End of Feudalism

It has often been said that artillery knocked down the castles of the feudal lords and created unified states. In a general way this is true, but the process was actually a good deal more complicated. The early cannons were enormously expensive. That, in itself, greatly reduced the weight most nobles could throw around. In fact, few of them could afford cannons. If they persisted in defying the king, royal artillery demolished their castles. Meanwhile, the big mercantile towns could afford cannons, which decisively tilted the scales in the age-old conflict between the aristocracy and the bourgeoisie.

No matter how expensive cannons were, they weren't nearly as expensive as the new cannon-proof fortifications pioneered in Italy. The new fortresses required more digging and more masonry than the old castles. They also required many guns. Siege engines before gunpowder were not much more than annoyances to the garrison of a strong castle. Catapults and trebuchets could throw boulders over the walls, but the besieger usually ran out of food before they could breach a wall. Battering rams were not much more effective. Cannons were a different story. To counter cannons, you had to have cannons yourself. Expense upon expense. Ventilation had never been a crying need in the old castles, but gunsmoke is not oxygen. Gun positions had to be designed to dissipate it, which also cost money.

And because guns had eliminated all those minor nobles, attacking armies were different. No longer were wars fought between Lord Notsohot and the Earl of Nothingmuch. Now, the armies belonged to the king of Spain, the king of France, the Holy Roman emperor, and the sultan of the Ottoman Empire. Or at least, by the Republic of Venice, the Ottoman Empire's rival for control of the Mediterranean. This meant the fortresses had to be big. Very big.

There were still fortresses, and there were still sieges. There would be for centuries. But the little guys had been eliminated. Cannons had not blasted the aristocracy out of existence. But if most aristocrats tried to wage war on their own initiative, they'd be bankrupt before they began.

Great as was its effect on land warfare, gunpowder was responsible for an even more far-reaching turning point when used in naval warfare.

7

Command of the Sea: War Under Sail

John Paul Jones's *Bonhomme Richard* (right) fights the British ship *Serapis* off the English coast on September 23, 1779. Sailing warships ruled the seas for three and a half centuries. (U.S. Marine Corps photo of painting by William Strickland, National Archives.)

The Timeless Galley

IF ONE OF THE ROMAN sailors Julius Caesar sent to eliminate Mediterranean pirates were to come back in the sixteenth century and enlist in Andrea Doria's Genoese fleet, he wouldn't find much that was new.

Neither, for that matter, would Odysseus. The Mediterranean warship was still the galley—a long, low, lightly built ship propelled by oars. The galley, like the Greek trireme of 2,000 years earlier, still had a ram on its bow, and Doria's mariners, like Caesar's sailors, relied heavily on boarding enemy ships. The biggest change was that the ram was no longer the most important weapon on the Mediterranean warship. It had been replaced by three or four cannons mounted on the bow. But the tactics were the same: charge the enemy head-on. Instead of crashing into his ship, blast it with your cannons. Then, as always, close with the enemy, board his ship, and fight him hand to hand.

Some galleys still had catapults, and some galley captains were as ingenious as the ancients in using them. One tactic was throwing soft soap in ceramic pots at enemy ships to make their decks slippery. Another tactic was to fill those pots with poisonous snakes to distract the enemy crew. However, probably neither of these tactics were used immediately before boarding. In the seventh, eighth, and ninth centuries, the Byzantine Empire had a really terrifying naval weapon: Greek fire. This was a secret mixture (it's still secret) that was pumped through nozzles in the front of a galley. It ignited as soon as it hit water, which in most cases would have been almost immediately after it left the nozzle. Enemy ships were drenched with liquid fire. The Byzantines used Greek fire to destroy an Arab fleet in the seventh century—the first serious defeat the Arabs had suffered up to that time—and Russian and other fleets after that. Most analysts think the mixture contained quicklime, which develops a great deal of heat when it encounters water and could have ignited the other components. The other components were probably naphtha and other inflammable substances. Later, during the Crusades, the Muslims shot containers filled with naphtha or crude oil at the crusaders; shortly before launching the containers with their mechanical artillery, they ignited the contents. It wasn't the original Greek fire, however, because the formula had been lost by that time.

Cannons more than made up for the lack of Greek fire in the sixteenth century. The galley was still a formidable warship. In battle, the mast was taken down and propulsion was strictly by oars. Compared to a sailing ship, the oared galley was dazzlingly maneuverable: it could pivot, go backward, and accelerate instantly. It was lightly built, so the oarsmen—each rower was calculated to produce one-eighth of a horsepower—could give it speed. The recoil of the guns on the bow was absorbed by the full length of the ship. With the sides lined with oars and the stern occupied by two steersmen operating huge steering oars, the bow was the only place the guns could go.

The galley, however, had some weaknesses. The sides were low, because rowers couldn't be too far above the water. That, and the ship's relative fragility, meant that sailing during the winter in the Mediterranean was dangerous. Sailing in the Atlantic in any season was simply madness. The huge crew needed to operate a galley and the lack of cargo space meant that the ship had to put ashore frequently for food and water. But for fighting on the inland sea, the galley had been the capital ship for thousands of years.

Some sailors did venture out on the ocean. Julius Caesar reported that the Gauls had ships with leather sails that could sail against the wind, something Mediterranean sailors had not been able to accomplish at that time. The Gaulish ships were not warships, though. Their high freeboard, which was necessary to cope with ocean waves, prevented the use of oars. And without oars, they could not maneuver quickly enough to use ramming tactics. Sailors called these merchantmen, and similar commercial ships in the Mediterranean, "round ships." Their beam was much greater in proportion to their length than in a galley.

Early Freighters

Freighters developed more rapidly than warships, both on the Mediterranean and on the northern seas, to keep up with the increase in international trade. It was cheaper to move merchandise by water than by land, especially if the water was open sea and the transportation was provided by sailing ship. River boats provided cheaper transportation than wagons, and sailing ships were cheaper to use than river boats. During the height of the Roman Empire, sea trade got a big boost when the Romans eliminated pirates. Roman merchantmen were carvel-built (planking met edge to edge), had keels, and a single sail. They were about three and a half or four times longer than they were wide. They were partially decked and had a single square sail.[1] At the same time, northern nations were also building round ships. These were clinker-built (the edges of the planking overlapped), and, as the Gauls demonstrated, the northern sailors learned to handle their single sail so that their ships could more or less proceed against the wind.

In the ninth century, Greek ships began using the triangular lateen sail, which made it possible to beat into the wind. They probably got the new sail from the Arabs, who may have invented it or borrowed it from Persia or India.[2] By the eleventh century, the lateen sail was common in the Mediterranean. About the same time, Mediterranean ships began using the stern-post rudder, which was a far more precise way of steering than the clumsy steering oars previously used.[3] The compass may have been invented in China, but Italian sailors had been using a crude form of it before A.D. 1000. The original was a needle floated on straw. But by 1269 it was pivoted and set in an instrument that moderns would recognize as a compass.[4]

These are but a few of a whole series of improvements in ships and navigation between 1280 and 1330. Merchant ships became larger and stouter and could sail in all kinds of weather, even on the ocean.[5] With increased navigation came increased opportunities for pirates. The cargo ships couldn't ram, and they didn't have a crew of rowers/soldiers like galleys. The answer was supplied by the merchants of Venice and Genoa: crossbows. There followed a boom in crossbow manufacture and a demand for Italian crossbowmen, especially Genoese, as mercenaries from all over Europe.

While all this was going on in the Mediterranean, shipping in the North and Baltic Seas was also developing. Northern sailors began shipping grain and other merchandise in a clinker-built, tubby ship called a "cog." The cog had a single square sail like the old round ship that all northern seamen, including the Scandinavians, had been using, but it was bigger and had a stern-post rudder. The northerners began trading with more southern ports along the Atlantic coast and into the Mediterranean. At the same time, the Mediterranean sailors ventured into the Atlantic and began trading as far north as Iceland. Northern and southern sailors met and swapped ideas.

The result was two new ships: the caravel and the carrick. The caravel had two or three masts holding lateen sails. It had high sides and "castles" over the bow and stern. It was longer and slimmer than the round ship and much faster. The carrick was bigger than a caravel. It had three masts, two supporting square sails, and a lateen sail. And both ships had stern-post rudders. Each of the masts had a crow's nest where crossbow archers could be stationed.[6]

Early on, both caravels and carricks began mounting cannons. At first, the guns were put on the castles, but the heavy guns there made the ships unstable. Later, heavy guns were placed on the lower deck and light breech-loaders fired from swivels on the castles. At the beginning of the sixteenth century, shipbuilders cut gun ports in the hull that could be closed in case of high seas. Gun ports appeared on both caravels and carricks. King John of Portugal was fond of caravels. He used them to mount heavy guns whose shot skimmed over the water and smashed through enemy hulls where they were most vulnerable. Although caravels were small, they could carry as many as forty cannons. Most of them, though, made do with fifteen heavier cannons. Carricks were much more heavily gunned. The English carrick, *Harry Grace a Dieu*, launched in 1514, carried 186 guns.[7]

Dar es Islam

Transoceanic commerce was by no means a monopoly of Europeans. In the first half of the fifteenth century, China had the world's most powerful navy. Chinese ships sailed to what are now the islands of Indonesia, Ceylon, India, and even Arabia and Africa. The enormous warships, more than four times as long as Christopher Columbus's *Santa Maria*, awed the local potentates who readily agreed to buy whatever the strangers from the East offered.[8] The Chinese officials called the payments for their goods tribute and considered the states their fleet encountered dependencies. But Chinese merchants, working with corrupt court officials, aroused the ire of the Confucians, who revolted and overthrew the government. Back in power, the Confucians, who disliked both merchants and military men, abolished the navy and ended foreign trade.

Indians, Persians, Arabs, Malays, and Africans had been engaged in somewhat less bellicose commerce for centuries before the birth of Mohammed. Their lateen-rigged *dhows* shuttled across the Indian Ocean ahead of the monsoon and other seasonal winds. The round trip took several months to complete, because the winds blew in one direction part of the year, then changed direction for the next part. It was slow, but transportation was cheap. When Islam spread over the Near East, the Middle East, India, and the Indies, *Dar es Islam*, the Land of Islam, grew rich. Even the Crusades helped. The incursion of Christian warriors was certainly a problem for the Muslim rulers of lands on the eastern Mediterranean, but it also opened new markets. The Christians became acquainted with and addicted to the goods of the East—spices, silks, ivory, gold, jewels, and other items not easily available in west European states just emerging from barbarism. The addiction continued after the crusaders had been pushed out of Palestine and Syria.

Some of those goods from China, India, and the Indies traveled west in caravans plodding along the famous Silk Road. But after the breakup of the Mongol empire, that route became much less popular. Bandits robbed the caravans or demanded payment for protection. Local rulers, hardly less rapacious, charged tolls for safe passage. But the ocean had no rulers, and pirate galleys operated close to the shore. Dhows scudding across the Indian Ocean from India to Africa seldom came close to the shore. Eastern merchandise landed in Arabia or Africa was usually loaded on other ships and taken up the Red Sea. The sultan of Egypt controlled much of this trade, and Egypt practically monopolized the trade with Europe. The sultan was a strong influence on the foreign policy of Venice.

Venice, Genoa, and other European ports grew rich from this trade with the Orient, but not nearly as rich as the rulers of Islam. Civilization flourished under the crescent. Barbaric Turkish nations were drawn into Islamic civilization. Arab traders brought their culture to the Indies, and Malay sailors spread it farther in the South Pacific. Art, architecture, poetry, and philosophy in Dar es Islam were light years ahead of anything available in western Christendom. Christendom surpassed Dar es Islam in only two ways: the development of sailing ships and of guns.

The Land of War

To Muslims, Europe west of the Balkans was the Land of War—that is, an area where they would go only to make war. The name was appropriate in another way. War had shaped Europe more than any other continent on the globe. The history of Europe, someone said, is written in blood. The ancient Greek city-states were continually at war with each other. They got so good at it that Greek mercenaries were in demand in Egypt, Persia, Carthage, and about every civilized country west of India. Then Alexander the Great took his army to India. The Romans, using Greek tactics until they invented a superior system, warred with all of their neighbors for centuries and created the Roman Empire.

In the second century A.D., the tide turned. For the next three centuries, the Romans were busy trying to repel invaders. In the fifth century, the western Roman Empire collapsed. In its place was a welter of barbarian tribes and war bands. Goths and Huns, Lombards and Alans, Marcomanni and Alemanni rampaged back and forth across western Europe and almost destroyed civilization. Before the devastated continent could recover, Avars, Bulgars, and Slavs attacked from the east, Vikings from the north, and Moors from the south. The Vikings had no sooner been half-tamed when their sons, the Normans, attacked England and Sicily almost simultaneously. Between invasions, local magnates, most of them descended from chiefs of barbarian war bands, were fighting each other and armed merchants were trying to establish the independence of their towns. For most of this time, there was no central government with real control in what was to become France and Germany, Spain was divided between warring Christian and Muslim principalities, and England had a central government that seemed to be engaged in an endless series of civil wars. For a while, a warrior class, the knights, dominated fighting in Europe. Unlike their counterparts in Persia and Japan (most of the time), they never had a monopoly on power. And from the thirteenth century on, infantry had been defeating cavalry quite frequently. There was no limit to the market for new weapons, since the market was not confined to a single class, and there was a continual demand for more and better weapons.

Geography had much to do with European disunity. Basically, Europe is a massive peninsula on the northwest corner of Asia. Into it were crowded many diverse peoples: Finns, Balts, Slavs, Germans, Greeks, Latins, Celts, Magyars, Turks, and others. Each nation found its niche or niches. The peninsula called Europe is a mass of other peninsulas—Scandinavian, Iberian, Italian, and Balkan, among them— and islands, from Iceland to the Greek archipelago. Mountain chains cut the southern peninsulas off from the rest of Europe. Other ranges, like the Caucasus and the Urals, separate Europe from Asia. All of these features promoted independence, and consequently quarrels.

Another thing geography promoted was seafaring. Few places in western Europe were far from the sea. This was not the case in Egypt, Persia, China, or even India. The situation aided the development of ships as well as weapons. And the combination of ships and weapons changed the world.

The Crusades showed the Europeans that there was no way to expand to the East. Steppes and deserts, guarded by horse archers who had been developing their tactics and strategy for centuries, blocked the way. And Europeans badly wanted to get to the East. The East was where all those wonderful luxury goods came from. To get them, they had to pay exorbitant prices, most of which went to infidel potentates like the sultan of Egypt or the "unspeakable Turk." Religion was another motive. It was known that off in the depths of Asia or Africa there were Christian kingdoms. Pope Innocent IV had sent delegates to the great khan to open communications with him and with the Nestorian Christians who were among his subjects. Coptic Christian monks from Ethiopia had visited Portugal.[9] From this solid evidence grew the rumor of Prester John, a Christian king in the Orient, who might help his coreligionists by attacking the rear of the Muslims. If true, it would be a great benefit. The Turks were slowly advancing into Europe, and Turkish ships were gaining control of the Mediterranean.

War against the Moorish Muslims by the Christians had been going on for centuries on the Iberian Peninsula. Partly to take the war to the Moors' homeland, and partly to explore new markets, the Spanish and Portuguese put to sea. The Spanish discovered the Canary Islands and the Portuguese the Azores. The Spanish were still fighting the Moors in Spain when the Portuguese had freed their own country. The Portuguese then concentrated on Africa. The Portuguese captured the Moorish port of Ceuta, which was the northern terminus of the trans-Sahara trade. They gained a new appreciation of the riches to be gained in Africa and decided to concentrate on trade with the unknown kingdoms of the dark continent while they tried to sail around it and reach the Far East. With royal patronage, the Portuguese sailors began a methodical exploration of the African coast. Every year, they would explore 100 leagues of coastline, make treaties with native rulers, and establish trading posts.

The farther south the Portuguese ventured, the farther they got from civilization. They also passed by the great kingdoms of West Africa. They signed treaties with the lesser kings in the Congo basin. Finally, they approached the Cape of Good Hope, the land of the Bushmen. Here, there were no kings and kingdoms, nothing but families of hunter-gatherers.

A sailor known as Old Álvaro wrote, "The inhabitants of this country are brown. All they eat is the flesh of seals, whales and gazelles and the roots of herbs. They are dressed in skins and wear sheaths over their private parts."

The ships of Bartolomeu Días got around the stormy cape and proceeded north. Suddenly, it seemed, they were back in civilization. In place of the bands of hunters there were port cities with stone piers and multistoried buildings. The people were of many races: Negro,

Indian, and of mixed Arab, Persian, and Negro blood. They were also all Muslims. The Portuguese were the first Christians they had ever seen. Most of the inhabitants took Días's men for Arabs or Turks. Días fought some skirmishes with the rulers of some East African city-states and made allies of others. He crossed the Indian Ocean and landed at Calicut, a cosmopolitan port that conducted a lively trade with Persia, Arabia, and Africa and had been an important stop for the great Chinese "treasure ships" a century before. The Hindu rajah, who had never seen Europeans before, welcomed them as he welcomed all foreign merchants. The resident Muslims, however, did not welcome the "Franks" (all west Europeans were "Franks" to them).[10] They saw Christian entry into the Indian Ocean trade as a major threat. They incited the rajah against the newcomers and Días had to leave.

The Portuguese did not give up, however. They sent new expeditions to India under Pedro Álvares Cabral (who accidentally discovered Brazil en route) and Vasco da Gama. The Portuguese helped the rajah of Cochin, who was warring against the rajah of Calicut. The Portuguese followed up their victory by seizing key points on the Indian Ocean coast and sinking any Muslim ships they could find.

The Empire Strikes Back

Portuguese convoys began traveling to Africa and India routinely. Muslim merchants felt the pinch, as did the sultan of Egypt and his trading partners in Venice. The Venetians joined the Arab and Persian merchants in urging the sultan to do something. The sultan went to his archrival across the Mediterranean, the ruler of the Ottoman Empire. The Egyptian and the Turk agreed to combine their fleets and send an expedition down the Red Sea that would settle the hash of the Portuguese forever. They launched a huge fleet of galleys from Jeddah on the Red Sea and sailed into the Indian Ocean.

Meanwhile, Francisco de Almeida, named viceroy of the Indies by the king of Portugal, had arrived in the Indian Ocean and sent ships to various ports of interest to Portugal. He heard that some Muslim ships were located at the port of Diu on the Indian coast. He sent his son, Lorenço with a few caravels to scout out the situation. Lorenço found, instead of a few Muslim ships, the enormous fleet of Turco-Egyptian galleys. The Muslims trapped him and attacked. Lorenço was killed. The Turks skinned his body, stuffed the skin with straw, and sent it back to Constantinople. Then they sailed back to Jeddah before Almeida could gather his fleet.

Encouraged by their victory, the Turco-Egyptians returned two years later with even more galleys and 15,000 marines for boarding the Christian ships.

Two Arab chroniclers record what happened, rather skimpily and none too accurately:

"Kansuh al-Ghawri came to power [as sultan of Egypt]. He dispatched a mighty fleet to fight the Frank, its commander being Husain Kurdi. Entering India, he stopped at Diu. The expedition fell in the year 13 [in the Muslim calendar]. It had an engagement with the Frank, but was defeated and returned to the Arabian coast. This was the first appearance of the Franks, may God curse them, in the [Indian] Ocean seizing [Muslim shipping]," wrote Ba Fakhi al-Shiri.[11]

Shanbal, another Arab historian, gave a little more detail:

In this year [1508–1509] the Frank took Dabul, looting and burning it. In this year also, the Frank made an expedition against Gujerat and attacked Diu. The Emir Husain, who was at that time in Diu fighting the Holy War, went forth to meet him, and they fought an engagement at sea beyond the port. Many on the Frankish side were slain, but eventually the Franks prevailed over the Muslims, and there befell a great slaughter among Emir Husain's soldiers, about 600 men, while the survivors fled to Diu. Nor did he [the Frank] depart until they had paid him much money.[12]

For the Muslims, the Battle of Diu was much worse than either chronicler makes it sound. It was worse than a disaster. It marked a turning point both in naval warfare and in world history. It, and three more battles in the Indian Ocean, all of which were mere replays of Diu, ended forever Muslim control of Oriental seas. It let Europe strategically outflank Dar es Islam and started the lands of Islam on their long slide into what a few years ago was fashionable to call the Third World. Here's what happened:

When Hussein al-Kurdi returned with his enlarged fleet, Almeida was still in the Indian Ocean. Burning for revenge, he was ready for battle. He had seventeen ships, and he led them all to Diu as soon as he heard that the Muslims were there.

The Muslims had 200 galleys. They had thousands of rowers, all of whom were fighting men, in addition to the marines. The marines carried bows and matchlocks besides the spears and swords available to all the crews. They had grappling irons for seizing ships and fire pots to drop on their enemies' decks. Each galley had three cannons over the bow, all loaded and ready to blast the Portuguese. As soon as his scouts saw sails on the horizon, Hussein led his galleys out of the harbor. The ocean was rougher than the Red Sea. The rowers had trouble making consistent speed, and the captains found it harder to form a straight line as they charged the Christians.

The Portuguese didn't charge. As the Egyptians and Turks approached, Almeida's ships turned and fired broadsides. The Portuguese guns were heavier than the Muslims' and they had longer range. And there were more of them on the 17 caravels and carricks than on the 200 galleys. The Portuguese cannon balls ripped through the banks of Muslim oarsmen, leaving masses of gore, mangled bodies, and body parts. They blasted holes in the flimsy galley hulls, sinking galley after galley. The Turco-Egyptian fleet was almost annihilated. As an Indian writer put it, "Courage availed nothing against artillery, and their fragile craft were sunk in batches."[13]

The Oriental trade, that great source of wealth to Dar es Islam, was gone. The heaviest loss was to Egypt, once the great power in the eastern Mediterranean, the Red Sea, and the east coast of Africa. Eight years after Diu, the Ottoman Turks conquered Egypt. Turkey had resources other than the Oriental trade. But after the sixteenth century, it slowly declined in wealth and power. So did Persia, but its decline was swifter.

In naval warfare, the change was abrupt. The oceangoing sailing ship, bristling with heavy guns, had completely replaced the galley. It would remain the capital ship of all the world's navies until the coming of steam, and even then, the change was merely a matter of propulsion. There was one more great galley battle, again between Turks and Christians, in 1571 at Lepanto. But even in that, the Christian commander, Don Juan of Austria, used galleasses—large sailing ships capable of firing broadsides—to break up the Muslim formation. Lepanto was a victory hailed throughout Christendom, but it was also the last gasp of an obsolete form of war.

8

The Bloody Birth
of Standing Armies

Seventeenth century musketeer. Soldiers like this fought in the wars of religion and later in regular armies.

War and Private Enterprise

NORTHERN ITALY, as we have seen, eliminated the scourge of "free companies" by hiring mercenaries in very small units and paying them through taxes on the citizens. What those cities got was very close to national armies. It was quite a while before other countries in western Europe could follow the Italian example. That's because money was scarce north of the Alps. During the Dark Ages, most of what commerce existed was conducted through barter. Much of the fighting in those days was done by feudal levies. Campaigns were short because vassals were bound to serve their lords for only a limited time. After that, the lord, usually the king, had to pay soldiers. Kings got the money to hire mercenaries by allowing knights to pay *scutage*, a "shield tax," in lieu of military service. Since a mercenary soldier was unlikely to fight for twenty pounds of flour, a side of beef, or a flitch of bacon, scutage was paid in money. With the money they received from the scutage, the kings paid soldiers. The soldiers bought what they needed. The peasants and villagers who sold goods to the soldiers used the money they received to buy other things. The whole process stimulated commerce and industry. War, therefore, became a major incentive to move the economy of northern Europe away from the barter system.

As the Dark Ages moved into the Middle Ages, more and more fighting was done by mercenaries. Because money in northern Europe was scarce even for kings, mercenaries were sometimes unpaid. When they didn't get their pay, the troops "lived off the land," which is a euphemistic way of saying they robbed the peasants. There was a great deal of that during the Hundred Years' War, both by French and English mercenaries, but the English mercenaries were by far the worst. The war was fought exclusively in France, so the devastation of the French countryside didn't particularly bother the English kings. It did, however, bother the French kings, and it bothered the French nobility, townsmen, and peasants even more. The king might give the mercenaries regular salaries as the Italian city-states did, but where would he get the money? France had nothing like the sophisticated economy of those commercial centers across the Alps.

Levying taxes was not easy for a king in medieval times. That was long before the Divine Right of Kings doctrine was accepted. Although in theory the king was overlord of all the lords, he was usually only the strongest of a mob of powerful nobles. Sometimes, he wasn't even that. The nobles, secular and ecclesiastic, had a voice in the government. The English Parliament, the French Estates General, and the Spanish Cortés were not originally rubber-stamp organizations. Later, the monarchs tried to make them just that. The parliamentarians responded by cutting off the heads of the English and French kings, while Napoléon Bonaparte disposed of the Divine Right of the Spanish monarch. The Hundred Years' War made it easier for the king of France to tax his subjects. The subjects greatly preferred paying taxes to being robbed, burned out, and murdered. The king was able to levy more taxes, and his subjects were willing to pay them to get the military brigands off their backs. Toward the end of the war, the French were able to hire a regular army. That, and their new, mobile siege artillery, drove the English out.

The Spanish government decided to imitate the French and hire a regular army. Spain had nothing like the resources of France, so Philip II tried unsuccessfully to raise cash through taxes. Europe, however, wasn't a vast unified area with a single government like China. When Philip taxed cannon makers too much, they closed their shops and moved to another country.[1] Instead of making the Crown richer, excessive taxes made Spain poorer. But there was another way to raise an army.

What Spain had was gold and silver from its American colonies. With it, the Spanish were able to hire a fine army—the finest in Europe. The Spanish kings spent their money on soldiers' salaries and on weapons produced in other European countries instead of on promoting domestic industry, including an arms industry. The doge of Venice could probably have told the king of Spain the danger of creating an army this way, but it probably wouldn't have mattered since European kings were not very savvy about economics. All that precious metal from the New World caused runaway inflation in Spain. Real buying power dropped, and the king had to borrow from northern Italian and southern German bankers. Then King Philip II went bankrupt. Spain's bankruptcy ruined the Fuggers, a banking family that had been the financial powerhouse of central Europe, and created a financial panic all through Europe. Worse, it caused a mutiny among Spanish troops in the Netherlands, who were fighting the Calvinist-dominated United Provinces. Instead of simply refusing to fight, the usual recourse of unpaid soldiers, the mutinous troops sacked Antwerp and massacred many of its citizens. Antwerp had been the richest city in the Low Countries. It never fully recovered. People with money to invest in commercial and industrial enterprises took it to Amsterdam, which was controlled by Philip's enemies.[2]

The king of England didn't get a chance to create an army like his French and Spanish neighbors. The thousands of unemployed soldiers and the lords who had been their officers in the Hundred Years' War plunged into the Wars of the Roses. Those wars lasted thirty-two years and gave England a taste of what France had been going through for the last century.[3] The English didn't like it. They disliked it so much that they were willing to put up with the despotic Tudors, who introduced the Divine Right doctrine to England. The Tudors at least provided peace and eliminated what has been called "bastard feudalism"—that is, rule by an aristocracy not based on ancient rights and obligations, but on the ability to hire mercenaries and maintain private armies. It wasn't until Henry Tudor became Henry VII that England even started to develop a regular military establishment. And then it consisted of nothing more than the king's bodyguard. Henry VII was a notorious skinflint. Not so his son, Henry VIII. Henry had not been raised to be a king. That was to be his older brother's job. Henry devoted himself to sports, especially martial sports like archery and jousting. When his brother, Arthur, died and Henry found himself about to be king, his most ardent desire was to make England a great military power. He hired gun founders and built mighty warships like *Harry Grace á Dieu*. He hired Burgundian cavalry and German infantry, Spanish handgunners, and Italian artillerymen. He revived the ancient militia laws and built shipyards and coastal forts. Being king with Divine Right, he was able to tax his subjects. And when he established his own church, he gained more cash by looting

Roman Catholic monasteries. But for all that, Henry never created a standing army.

Henry's son, Edward VI, organized a "gendarmery" modeled after the similar organization that was France's regular army. But Edward VI died very young, and so did the gendarmery. His half sister, Mary, brought the militia standards up to date and did much to increase the efficiency of the militia, but she did not organize a regular army. During her reign and that of her half sister, Elizabeth, the lord lieutenant of a county chose junior officers from the gentry, who then drafted common soldiers from the militia. Military service was not popular with many of the militiamen, and their morale tended to be low.

An alternative method of raising an army was a system that somewhat resembled the organization of an army of mercenaries. It more strongly resembled the creation of privateers, who were the basis of much sixteenth- and seventeenth-century naval strength, particularly in England. Under the alternative system, the queen would appoint a noble as a captain and commission him to raise a company of volunteers. He would receive the money necessary to equip, supply, and pay the troops. How well he equipped, fed, and paid them pretty much depended on his conscience. A common practice among captains was failing to enlist a full company but drawing pay for soldiers who did not exist. It got so bad that the Queen's Privy Council ruled that if the strength of a company of 150 infantry dropped below 100 or of a company of 100 cavalry fell below 60, the captain would be dismissed.[4]

Privateers were sometimes issued letters of "marque and reprisal," which authorized them to attack the ships of any enemy with which their country happened to be at war. In the sixteenth century, though, seamen, especially English seamen, were actually part-time pirates rather than authentic privateers.[5] A ship's captain would sell shares in his expedition, then he would put to sea to make money any way he could. It might be entirely legitimate trading or it might be smuggling, slave trading, or involve, as an old ballad put it, being "a robber upon the high seas." Queen Elizabeth was a heavy investor in these maritime ventures, and she was an acute businesswoman who knew which captains to invest in. She did well. So did the captains. Such naval heroes as Francis Drake and Martin Frobisher were actually pirate admirals who commanded more ships than the Royal Navy. At the battle with the Spanish Armada, Drake alone had as many ships as the entire Royal Navy.[6]

The Wars of Religion

This period, during which the west Europeans were coalescing into nation-states and creating national armies, was also the time of the Protestant Reformation and the great wars of religion. The Albigensian Crusade and the later Hussite Wars, previous wars of religion, were nothing like the wars initiated by the Reformation. In these early religious wars, the fighters were, as in all such conflicts, cruel and ruthless. But the fighting was nowhere as widespread as in the religious wars of the sixteenth and seventeenth centuries. All of Latin Christendom was involved. What emerged from this mass of mayhem was an all-European military system that was to dominate the world for the next two centuries.

The first of this cycle of religious wars began in Germany, where Martin Luther had published his ninety-five theses in 1517. Soon after that, fighting broke out between Lutherans, Catholics, and a variety of radical religious sects. These battles did not involve regular armies, because there were no regular armies in the German states at this time. They were bloody and vicious but broke no new ground in the art of war. The fighting spread to Switzerland, where Calvinists, rather than Lutherans, were the dominant Protestant group. The fighting was less chaotic than in Germany, because each Swiss canton had a highly trained militia that it rented out to rulers of neighboring countries. A generation later, Calvinists in Scotland overthrew their queen, Mary, who fled to England. Religious differences further inflamed the continual feuding among the Scottish clans.

Calvinists and Catholics also clashed in France. Here, the core of each side was composed of regular soldiers. The fighting was prolonged and serious. There were to be nine Huguenot wars. One new tactic that was used extensively in those wars was the *caracole*, a maneuver to allow cavalry to charge a wall of pikes. The infantry pike, eighteen to twenty feet long, had a greater reach than any lance a horseman could use. So cavalry abandoned the lance and adopted the wheel lock pistol. Each cavalryman would carry two, sometimes four pistols. The cavalry would form a column and ride up to the pike-men. Each man in the front rank would discharge his pistols just out of pike range and gallop back to the rear of the column, reloading as he did so. The idea was to bring continuous fire on a selected point on a line of pikemen. Use of the caracole prompted generals to increase the proportion of musketeers in their armies, since muskets greatly outranged pistols.

While this fighting was going on, the French were also fighting the Holy Roman Empire, which was also involved in the German reli-gious wars and in its long-running feud with the Ottoman Empire. The Holy Roman Emperor was a Hapsburg. So was Philip II of Spain, son of Emperor Charles V. The Bourbon dynasty in France was an ancient enemy of the Hapsburgs, and beating the Hapsburgs overrode all other priorities, including religion. To harass the Hapsburgs the Bourbons allied themselves to the Muslim Ottoman Turks.

The top priority of Hapsburg Philip II of Spain was not beating the Bourbons but rooting out heresy. This brought him into conflict with the Calvinist burghers in the rich commercial cities of the Low Coun-tries. The Low Countries, modern Belgium, Luxembourg, and the Netherlands, were the native land of his father, Charles V, and Philip inherited them along with Spain. Philip II, who sent the Armada against England, is seldom a favorite of English writers and Protes-tant historians, but his fanaticism—if that's what it was (the same dis-ease affected his sister-in-law, Elizabeth I of England)—was not the sole cause of the longest religious war in history, what the Spanish called the Dutch War, and what the Dutch called the Eighty Years' War. The intense feelings generated by this period make it hard to find an objective account of what happened. This is most unfortunate, because in this gory era are found the seeds of much modern military history, including the formation of regular armies.

In the Low Countries, the merchants, bankers, and manufacturers thrived. The coastal cities had excellent harbors and they also sat on the end of the commercial river traffic from the interior of Europe. Dutch seamen were taking the place of the Portuguese as kingpins of the Far Eastern trade. The nobility, however, was becoming impoverished. It was similar to a situation that developed in Japan a century or two later, when the samurai, the highest social class, grew increasingly poorer, and the merchants, the lowest class, waxed wealthy. Calvinism, which at that time held that worldly wealth was a sign of God's favor, was enthusiastically received in the cities of the Low Countries. From this point of view, the nobility would seem to be definitely out of favor. Nevertheless, they, too, adopted Calvinism.

According to General J. F. C. Fuller, although "Charles V's attempts to stamp out heresy had led to religious grievances, it is unlikely that they would have led to open rebellion had not the bankrupt Dutch nobility seen in the Protestant cause the means to despoil the monasteries and other ecclesiastical foundations."[7] In 1559, Philip II's half sister, Margaret of Parma, the natural daughter of Charles V and regent of the Low Countries, decided to reorganize the monasteries and other church properties. The reorganization would deprive the nobles of patronage and preferment, thus leaving them with no way to repay their mounting debts. This would be a mortal blow to the greatest of the nobles, William, duke of Nassau and prince of Orange. William is known in English as William the Silent. He wasn't particularly taciturn: the nickname is a mistranslation of the Dutch word *schluwe*, meaning prudent or cunning. William organized the other nobles and the magnates of the cities and began a rebellion. Bands of Calvinists from England, France, Germany, and Switzerland swarmed into the Low Countries and sacked 400 churches.[8] Trevor Davies, an English historian, says, "Calvinism . . . was the Third International of the sixteenth century."[9] Philip hated heresy, but he had other worries, too. He feared that the Dutch Calvinists would unite with their French coreligionists and that they then might spread their doctrines to Spain.

Then the Catholics of Groningen, in the northeast corner of the Netherlands, rose against the rebels. Philip sent the duke of Alva to their aid. William the Prudent left the Low Countries to gather a mercenary army before Alva arrived. Alva was an excellent general and he was leading the superb Spanish infantry. But he was also a harsh, even bloodthirsty man. In the field, after William returned, Alva beat the rebels repeatedly, inflicting horrendous casualties. But the Dutch cities were strongly fortified and had elaborate water defenses.[10] The war, which Fuller calls a "war of pillage and of retaliation,"[11] dragged on with neither side gaining an advantage.

William the Silent was assassinated in 1584. Four years later, the rebellious United Provinces of the Netherlands appointed his son, Maurice of Nassau, captain general. Maurice was a far better general than his father, but he was opposed by Alessandro Farnese, the duke of Parma. Parma was the son of Margaret of Parma and another descendant of Charles V, like his cousin, Don Juan of Austria. More important, he was a better tactician than Alva, while he lacked Alva's ferocity. He could possibly have ended the war if King Philip hadn't called him off at critical moments—first to prepare the southern Low Countries to receive the Armada, second to invade France as part of the eternal Hapsburg-Bourbon feud.

Parma's maneuvers excite the admiration of military historians, but it was Maurice of Nassau who made the greatest contribution to the development of armies.

Maurice increased the proportion of musketeers to pikemen in his army. He thinned the line of musketeers so there were more guns on the firing line. He also reduced the size of units. Instead of a massive square of pikemen—the system the Swiss introduced and the Spanish adopted—Maurice organized his infantry into battalions of 550 men, then subdivided the battalions into companies and platoons. He established a chain of command from the commanding general to the lowest noncom. This made his army both more mobile and more responsive to command.[12] Working with the gun founders of Holland, a leading producer of iron cannons, he got lighter artillery. A student of the Latin classics, he had his men revive the Roman custom of creating earthen fortifications wherever they camped. For Maurice, the spade was as important as the musket. Most important, he drilled his men continuously. The matchlock musket, the basis of infantry firepower at that period, required forty-two distinct movements to load, fire, and fire a second shot. The many writers who say the matchlock musket replaced the longbow because it required less training obviously have had no experience with either weapon. Maurice drilled his men incessantly in the necessary movements. Each was executed on command. As a result, when his troops were in combat, no one missed a step in the process, and the volleys were delivered faster. He trained his musketeers in the countermarch: when each rank fired, it did an about-face and marched through the ranks behind it and prepared to fire a second volley. The pikemen not only marched in step—an essential for any phalanx—they coordinated opening their ranks to let the musketeers take shelter from a cavalry charge.

The continual drill let the Dutch general's troops act automatically, or at least, semiautomatically in battle. This sort of training, when new weapons appeared—the flintlock musket and the bayonet—made possible the incredible eighteenth-century style of warfare, where soldiers stood shoulder to shoulder and exchanged volleys with similar soldiers only a few yards away. The importance of weapons drill can be seen in the American Civil War, where many rifles were found on the battlefield to have been loaded as many as five times but not fired at all. In the heat of battle, soldiers apparently forgot to put a primer cap on the rifle after loading it, pulled the trigger, and reloaded the unfired weapon. If the troops had been drilled by Maurice, they never would have forgotten the primer cap. Close-order marching drill was also part of Maurice's regimen. It had obvious benefits for troops using seventeenth- and eighteenth- century volley-fire tactics. William H. McNeill, in his *Pursuit of Power*, suggests other benefits. He says rhythmic movements by groups of men promote solidarity and cooperation—esprit de corps. "By virtue of the dance, supplemented and eventually controlled by voice signals and commands, our ancestors elevated themselves to the pinnacle of the food chain, becoming the most formidable of the predators."[13]

The Eighty Years' War ended with an independent United Provinces approximately where the Netherlands is today, and the

Spanish Netherlands, which were about where the modern kingdom of Belgium is today. Maurice's influence spread far from the Netherlands. Graduates of his military academy—the first in Europe—took his ideas all over western Europe. They took an especially strong root in Sweden, which seriously affected the most climactic of the religious wars: the Thirty Years' War.

The Slaughter of Central Europe

The Thirty Years' War began in Bohemia and was fought mostly in Germany. It was the last and most horrendous war fought mostly by mercenaries, but it involved half the nations of western Europe: Denmark, Sweden, France, and Spain, as well as Bohemia and the divided Holy Roman Empire. In addition, there were mercenaries from Ireland, England, Scotland, Italy, Poland, and possibly Russia and Turkey as well. It began as a war of religion and ended as one of dynasties and nationalism. It was a war of religion in which Catholics fought on the nominally Protestant side and Protestants on the nominally Catholic side and atheists and possibly Muslims on both sides. It may have killed almost half the people of Germany and ended without a victor.[14]

In 1555, the warring princes of Germany had signed the Peace of Augsburg, which specified that each ruler could choose the religion of his subjects. This was a compromise that didn't satisfy anybody, and troubles continued, leading to the formation of two factions in the Holy Roman Empire: the Evangelical Union of the Lutheran princes and the Holy Catholic League of the Catholic princes. There was no organization for the Calvinists, who were growing in strength in Germany. The pot came to a boil when Ferdinand von Hapsburg was elected both Holy Roman Emperor and king of Bohemia to replace the dying Matthias von Hapsburg who was also both. Ferdinand, a truly fanatical Catholic, refused to endorse Matthias's promise to the Bohemians to stop giving royal land to the Catholic Church. A mob in Prague threw his delegates out of a window and took back their crown from the king-elect. They gave the crown to a Calvinist, Frederick, the Elector Palatinate and son-in-law of James I of England. Ferdinand, as leader of the empire, declared war on Bohemia.

At that time, almost no state in central Europe had a regular system of taxation, so there were few standing armies as in France and Spain. In fact, there was only one: Maximilian of Bavaria's. Rulers were expected to live on the proceeds from their estates. In case of war, the prince could levy a special tax to hire a mercenary army. The mercenary captains—most of them called themselves counts, whether they were or not—bid for contracts. The low bidder usually got the job. Since he usually didn't charge enough to pay or supply his soldiers, they supported themselves by living off the land. Later, national armies—the Swedish, the French, and the Spanish—got into the fight. But they, too, lived off the land. And because this was a war of religion, they didn't merely rob the civilian population. The warriors tortured and killed them and burned their towns. Thirty years of this literally burned out the heart of Europe.

In July 1620, Duke Maximilian, with his army and two imperial mercenary armies, invaded Bohemia and routed Frederick's troops. At the same time Philip III of Spain sent another army into the Palatinate, Frederick's home country. The Catholics seemed to have triumphed, so the Evangelical Union broke up. Ferdinand could have ended the war right then. But he believed, "Better to rule over a desert than a country full of heretics." The war went on. The Protestants hired Ernst von Mansfeld, a Calvinist mercenary, who quickly became known as "the Attila of Christendom."[15]

The Catholic League supported its own mercenary captain, Johan Tserclaes Tilly, who, though once beaten by Mansfeld, smashed all the other Protestant armies, then avenged himself on Mansfeld. Frederick fled, and Ferdinand unconstitutionally placed Maximilian on his throne and sold the estates of Frederick's followers at auction. To save the Protestant cause, the Lutheran king of Denmark entered the war. Denmark's national army was more powerful than any army in Germany, and it looked as if the course of the war had suddenly reversed. To add to Ferdinand's troubles, France declared war on the Holy Roman empire.

The Supreme Military Adventurer

At this point, the first of the two towering figures of the Thirty Years' War joined the game: Albrecht Eusebius Wenzeslaus von Waldstein, known to history as Count Wallenstein. Like the other towering figure, King Gustavus Adolphus of Sweden, Wallenstein was a genius. And like Gustavus, after a spectacular career, he accomplished nothing. And also like Gustavus, he was not a German. He was a Czech. Most historians writing in English sanctify Gustavus and demonize Wallenstein, while those writing in German view Wallenstein a national hero and Gustavus as a foreign interloper.[16]

Wallenstein had served the emperor before and was known as a highly competent officer who had twice beaten the Hungarians and captured a Bohemian wagenburg. He had purchased vast tracts of land at Ferdinand's auctions of Protestant estates and added them to his holdings in Moravia. He made a huge fortune on these properties by introducing such radical measures as building schools and roads in his villages, draining swamps, and introducing scientific agriculture. He hired experts from all over Europe to develop his properties and he gave his tenants a charter guaranteeing many rights and privileges, including religious freedom.[17] These reforms may have come from the goodness of Wallenstein's heart, but it is certain he knew they'd increase his wealth. With that wealth, he made loans to the emperor when that monarch was short of funds.

When Ferdinand could see nothing but disaster in his future, Wallenstein appeared again. Ferdinand hoped this rich vassal could provide enough money for him to hire new troops. Wallenstein did better than that. He offered to raise an army of 50,000 men. All the emperor would have to do would be to pay their salaries. Ferdinand was not sure that Wallenstein was qualified to lead an army, but 50,000 men was 50,000 men. He gladly accepted. Things were looking better. Another Huguenot war had broken out in France, so the French dropped their planned offensive against the empire.

Wallenstein took his army of untested farm boys to meet Mansfeld's veterans. To a neutral observer, this looked like the makings of a disaster. It was. Wallenstein's farm boys practically destroyed Mansfeld's mercenaries. The Czech entrepreneur had lavished as much care on the training of this army as he had on his estates. Wallenstein and his army chased the remnants of Mansfeld's army all across Europe until the old mercenary died in Dalmatia. The Czech then returned to deal with the Danes. He chased Christian of Denmark out of Germany and even off the Jutland Peninsula, the Danish mainland. On the island of Zeeland, Christian sued for peace.

Wallenstein marched along the Baltic coast, conquering everything he saw and recruiting soldiers everywhere. Soon, he had the largest army in Europe under his direct command. Wallenstein had established his own supply depots in Bohemia, but he and his army usually lived off the land. He requisitioned supplies from any town or duchy his troops occupied, whether friendly or enemy territory, which earned him no friends among the Catholic nobility. Neither did his admonitions to the emperor to go slow in attempting to convert the conquered Protestants.[18] Spanish officials wanted him to take his army west to help their troops in the Netherlands. Wallenstein had no care for the Netherlands. He continued east to secure the rest of Germany. The king of Spain threatened to cut off the subsidy of silver he had been sending Ferdinand. The emperor summoned up his courage and ordered Wallenstein to retire. To the relief of everyone in Vienna, he did.

His army, though, did not join the imperial forces. Many of them followed Wallenstein back to Bohemia. Others went, with Wallenstein's blessing, with his trusted lieutenant, Georg von Arnim, a Saxon Lutheran. Von Arnim joined the army of John George, the elector of Saxony, a Lutheran and a neutral. At this point, the second towering figure, Gustavus Adolphus, entered the fray.

The Snow King

Of the king of Sweden, one of his admirers said, "Gustavus was one of those born conquerors to whom peace is an ideal state, always for excellent reasons unattainable."[19] He was fighting against Denmark at the age of sixteen. When his father died in 1611, he became king. He later fought Russia and Poland while he developed a new type of army. Some of the hagiographers who have written about Gustavus say the new army was entirely the king's brainchild, giving him credit for innovations that had been made before he was born. For example, Richard A. Preston, Sidney F. Wise, and Herman O. Werner in their *Men in Arms*, say Gustavus reduced the weight of the musket rest and "[t]he musketeer was thus enabled to carry a sword. When attacked, he could now defend himself with the sword."[20] The famous Jacob de Gheyn drawings illustrating the Dutch musket drill in 1596—when Gustavus was two years old—plainly show Maurice's musketeers wearing swords.

What Gustavus actually did was take the ideas of Maurice of Nassau and improve on them. He made Maurice's firing line even thinner, so more muskets could fire in a single volley. He arranged his pikemen and musketeers in the "Swedish cross," a formation that made it easier for the pikemen to shelter the musketeers from cavalry charges and easier for the musketeers to support the pikemen with both direct fire and crossfire.[21] He made the existing paper cartridge standard for his troops, thus increasing their rate of fire. There is evidence that he replaced some of his matchlocks with wheel locks or muskets using the Baltic lock, an ancestor of the flintlock. He lightened the field artillery. Three men could manhandle his four pounder, a gun that accompanied his infantry and had a higher rate of fire than a musket. The four pounder was a short-range gun that usually fired only grape or cannister shot. The Swedish cannons were lighter, muskets were lighter, and pikes were lighter. As a result, the Swedish army not only had more firepower than any other, but it could also move faster.

In France, the astute Cardinal Richelieu saw that the Swedish king and his army might provide a way for the Bourbons to crush the Hapsburgs. Richelieu was, of course, a Roman Catholic, but like an earlier English cardinal named Wolsey, he put loyalty to his king before loyalty to his church. Richelieu's arguments appealed to Gustavus, as did his promise of financial support. The king wanted to advance the Protestant cause, and he was delighted at the prospect of a new world to conquer. The emperor's courtiers derided Gustavus as the "Snow

King" and predicted he would melt as he moved south. Their derision was short lived. Gustavus landed in Germany and commenced a campaign that was both startlingly new and depressingly old. What was new was the firepower and mobility of his army. Those qualities let him rout the army of Count Tilly at Breitenfeld in 1642.

Gottfried Heinrich von Pappenheim, Tilly's cavalry commander, charged the cavalry on the Swedish flank without orders. But Gustavus had placed companies of musketeers between his cavalry units. They blasted Pappenheim's horsemen, then the Swedish cavalry followed up and drove Pappenheim's "Black Cuirassiers" from the field. Tilly then swung his army around to attack the Swedish flank. It was a brilliant move and would have succeeded against any other army. But Gustavus was able to move his agile Swedes faster and enfilade Tilly's army with his own imperial artillery. The Swedes swept Tilly's men from the field.

The depressingly old part of the campaign was Gustavus's system of supply. Like Wallenstein, he had a base of supply and could move matériel from Sweden to the front. But like Wallenstein, he preferred to make the war pay for itself. That was true even in friendly territory. At one point, Gustavus wrote to Axel Oxenstierna, his prime minister, "We have been obliged to carry on the war *ex rapto*, with great injury and damage to our neighbors."[22] The lack of supplies from Sweden was not Oxenstierna's fault. It was the king's policy. Before Sweden entered the Thirty Years' War, the army took five-sixths of the government's total revenues. During the war, while Gustavus led a vastly larger army—40 percent of it was composed of non-Swedes—it took only one-sixth of the government's total revenues. Gustavus knew he couldn't afford the army he led in Germany, and he took steps to cut expenses drastically.

The hagiographers make Gustavus sound like the ideal Christian prince, waging war while avoiding harm to civilians. At the same time, they point to "the ravages of Wallenstein." Actually, each commander was unusual in having totally disciplined armies, and each hanged looters who looted without permission. But permission was given much more often than not. Both Gustavus and Wallenstein were princes of plunder. And because they exercised so much control over their troops, they are more to blame than such commanders as Tilly and Mansfeld.

Of Gustavus, C. V. Wedgwood says, "When for political or strategic reasons, he wished to ruin a country, his men, released from the customary restraint made up with interest for the opportunities they had been forced to miss."[23] The "opportunities" were primarily murder and rape. Robbery, called "requisitions," was standing operating procedure. According to Wedgwood, "He [Gustavus] plundered as no man had plundered before in that conflict, because he plundered to destroy the resources of his enemies. 'Your Grace would not recognize our poor Bavaria,' wrote Maximilian to his brother. 'Villages and convents have gone up in flames, priests, monks and burghers have been tortured and killed at Fuerstenfeld, at Diessen, at Benediktbeueren, in Ettal.'"[24] One does wonder how torturing and killing priests, monks, and burghers destroyed the resources of the enemy.

The Clash of Titans

At this, his darkest hour, Emperor Ferdinand wrote to Wallenstein, begging him for help. Wallenstein made the emperor wait. Meanwhile, he was writing to his old lieutenant, von Arnim, now employed by the elector of Saxony. He was trying to get the elector to join him in a Catholic-Protestant coalition to make peace and drive out the foreign troops. Von Arnim replied, "I have . . . urged peace on friend and foe." Unless German rulers make peace, he said, "our beloved Germany will fall a prey to foreign people and be a pitiable example to all the world."[25] Germans were starting to think of themselves as Germans first and Protestants or Catholic second. Former king Frederick of Bohemia refused Gustavus's offer to return him to the Palatinate as a Swedish vassal. Another German noble remembered with a shudder a verbal slip the Swedish king had made, "If I become emperor . . ."[26]

While Wallenstein was negotiating with the present emperor, he was playing a strange game with von Arnim. The Saxon general invaded Bohemia. Wallenstein called up his troops, but instead of fighting, he merely retreated before his former lieutenant. Ferdinand, now feeling threatened by both the Swedes and the Saxons, agreed to Wallenstein's terms. The Czech adventurer was to have full control of the army, with no interference from the emperor or his sons, and control of all peace negotiations with the right to make treaties. Spain would have no influence on the conduct of the war. When Ferdinand signed the agreement, Wallenstein counterattacked von Arnim—a counterattack that was as bloodless as the Saxon invasion. All the while, he was imploring von Arnim to persuade John George of Saxony to join the coalition. John George, though, was terrified of Gustavus. He was afraid to cross the Swede.

There was nothing bloodless about what Gustavus was doing. He was busily killing and burning out Bavaria. Duke Maximilian begged Wallenstein to help him in Bavaria. Instead, Wallenstein, using his newly acquired rank, ordered Maximilian to leave Bavaria and join him in an invasion of Saxony. He calculated that Gustavus would try to prevent the junction of the two armies in his rear. He was right. Gustavus left Bavaria, but he was too late to prevent Wallenstein from meeting Maximilian. Faced with 60,000 enemies, the Swedish king fell back to Nürnberg and built an entrenched camp. Wallenstein also entrenched. He built his camp where he could cut Swedish supply lines and attack Swedish foragers. Neither general wanted to attack an entrenched camp. Finally, Gustavus did attack and was repulsed. His men were starving, because they had already scorched the earth for miles around and Wallenstein was astride their supply routes. Gustavus retreated. Physically, Wallenstein's victory didn't mean much. Psychologically, it did. It robbed Gustavus of the aura of invincibility he had acquired since landing in Germany.

Gustavus went back to ravaging Bavaria. And, in spite of Maximilian's pleas, Wallenstein invaded Saxony. Again, Gustavus followed him. Wallenstein's friendship with von Arnim was no secret, and the Swedish king did not want the Czech adventurer to add the Saxons to his array of allies. Gustavus knew he had intimidated Elector John George, but Wallenstein, too, was intimidating. Wallenstein face to face was more intimidating than Gustavus many miles away. Gustavus may also have guessed Wallenstein's plan: secure the cooperation of the Saxons and move on to Denmark. King Christian of

Denmark was already afraid of Wallenstein, and he had no love at all for Gustavus Adolphus. With the help of the Danish navy, Wallenstein could cut Gustavus off from Sweden. The fast-moving Swedish army caught up with Wallenstein near Leipzig before he could meet the Saxons.

Gustavus built a fortified camp, but Wallenstein had no intention of attacking the entrenched Swedes with their enormous firepower. Then he made one of the few stupid decisions of his career. He thought that as the Swedes were entrenched, there would be no more movement for a while. He allowed the restless Pappenheim, a mercenary captain with his own private army, to attack Moritzburg while he waited in camp at Lützen. With Pappenheim gone, Wallenstein had only 8,000 infantry and 4,000 cavalry. Gustavus was not a general who would let an opportunity like that slip away. He led out 14,000 infantry and 5,000 cavalry.[27] When Wallenstein learned this, he sent General Ludwig Isolani and his Croatian light cavalry to delay the Swedes and sent a messenger to Pappenheim ordering him to return as soon as possible. Then he got his army in position to fight a battle.

Thick fog the next morning delayed the battle. At about 11 A.M. Gustavus ordered his artillery to open fire, then he drew his sword and led his cavalry against the imperial cavalry on his right. He pushed Wallenstein's horsemen back, but on the other flank, Bernard of Saxe-Weimar, leading the Swedish left, ran into trouble trying to take a fortified position on Wallenstein's right. Then the imperial cavalry, reinforced by musketeers—a trick Wallenstein had learned from Gustavus—pushed back Bernard. Wallenstein himself had taken charge there. The Swedish infantry, supported by Gustavus's quick-firing field artillery, crossed the sunken road that protected Wallenstein's center. Wallenstein, suffering from gout, had been directing the battle from a litter. But when he learned what was happening in the center, he climbed on his horse and led a counterattack. He was extremely lucky on this day: all his attendants were shot down, a shot from a cannon took off one of his spurs, and several musket balls hit his thick leather "buff coat," but he lost no blood.

Gustavus was not so lucky. Hearing of the reverse on his left, he took a regiment there to help Bernard. In the returning fog, he was hit by a musket ball and, apparently in shock, wandered off. Bernard suspected what had happened, but he told his men that the king had just gone to another sector.

About this time, Pappenheim returned. Basically a cavalryman, Pappenheim had returned as fast as possible, leaving his infantry to follow as best it could. He immediately delivered an old-fashioned hell-for-leather charge against the apparently triumphant Swedish right wing and drove it back in disorder. Then an unknown musketeer fired at a big horseman in black armor, ending the career of Gottfried Heinrich von Pappenheim, as brave and charismatic and as harebrained and reckless as any soldier who fought in this gory war. Pappenheim's men were shocked, as they had sworn loyalty to him alone. The Swedes took advantage of their confusion, rallied, and counterattacked. Meanwhile, Wallenstein and his infantry were still pushing back the Swedish center and Octavio Piccolomini, leading the imperial heavy cavalry, was again pushing back Bernard's troops.

Bernard came to a decision.

"Swedes!" he yelled above the din of battle, "They have killed the King!"[28] It was a gamble. If the Swedes reacted like Pappenheim's men, they were done for. They didn't. Burning for revenge, they furiously attacked the imperialists all along the line. Wallenstein's troops fell back. The Swedes recaptured the imperial guns they had taken and lost.

Then Pappenheim's infantry arrived in time to stop a rout. The imperialists streamed away from the field, but the Swedes were too tired to pursue. There was no celebration: they had found the king's body.

Aftermath

Wallenstein collected his troops and went into winter quarters. Along with reorganizing his army, he reopened negotiations with the Saxons. He proposed that John George join him in forcing the emperor to make peace and establish an empire that guaranteed religious freedom. Von Arnim approached Oxenstierna, the Swedish prime minister, with proposals for peace, but both he and Wallenstein were on record that the foreigners must leave Germany. It may have been that Oxenstierna was growing weary with his master's overseas adventures. The intrigues are all as foggy as the battlefield at Lützen.

What happened at the next battle is crystal clear, though. Wallenstein went after an army that combined Swedes and Saxons. In what British military commentator Basil H. Liddell Hart calls Wallenstein's "tactical masterpiece," he separated the Swedes from the Saxons then ambushed the Swedes. Surprised and surrounded, the commander of the Swedes surrendered on condition he and his officers go free and the men enlist in Wallenstein's army. The commander of the Swedish army, Count Heinrich Matthias Thurn, was not a Swede. He was another Czech, the leader of the mob that had thrown the emperor's delegates out of the window. Letting Thurn go was the last straw for the nobles in Vienna, although Wallenstein said he was such an inept commander a free Thurn was an asset for the imperial forces. But the nobles had also learned of Wallenstein's intrigues with the Saxons, and he had made no secret of his desire for religious tolerance. They may also have heard that Richelieu had offered Wallenstein a huge bribe and the crown of Bohemia if he would turn against the emperor. Wallenstein had curtly rejected the offer, but the nobles still didn't trust him. They sent a secret message to Piccolomini that Wallenstein was an outlaw and that they wanted him dead or alive, preferably the latter. On February 25, 1634, a band of imperial soldiers, led by Colonel Walter Butler, an Irishman, and Captain Walter Devereux, an Englishman, assassinated the Czech-born German patriot.

The death of Gustavus ended the possibility of a united Germany under Swedish tutelage. The death of Wallenstein ended the possibility of a united Germany. The Thirty Years' War dragged on for fourteen more years. The Swedish role decreased, but the French and Spanish roles grew. The boundaries of the German states changed very little and their religious preferences not at all.

One thing that did change was that the German princes had consolidated their power at the expense of the towns and instituted regular taxation. With this money, they hired soldiers and enlisted them for long terms. Governments adapted such unmedieval innovations as clocks, paper making, and printing to establish bureaucratic control over their armed forces, and generals learned how to command and coordinate large numbers of troops spread over great distances.[29]

All states, large and small, learned several lessons from the Thirty Years' War. From the army of Gustavus Adolphus, they learned the importance of firepower and mobility. From the Swedish king's emphasis on drill (which he learned from Maurice of Nassau), they learned how to increase their soldiers' efficiency. From the operations of Wallenstein, they learned the need to keep the military under civilian control, and that the more brilliant the military leader, the greater the need for civilian control.

The depredations by mercenary armies and by the national armies that followed them taught another lesson. In Germany, war against the civilian population almost destroyed commerce, industry, and even agriculture. Everyone got poorer, including governments, which could raise little tax revenue. Therefore, war against civilians should be avoided. All these lessons were well understood on the continent. Learning them in the British Isles took a little longer.

Almost all the nations of Europe developed standing armies under civilian control by the end of the Thirty Years' War. The one exception was the British. Another bloodletting, the civil wars of Britain and Ireland, created a standing army for them: Oliver Cromwell's "New-Model Army." And that, in turn, demonstrated forcefully the need for civilian control and the long-range harm of attacking civilians.

From the bloodbaths of the religious wars, Europe turned to a century of limited wars, with small professional armies battling each other safely away from civilian population centers. That ended with an explosion of total war—nation against nation instead of army against army.

9

Quiet Revolution: The Age of Limited Warfare

Hessian troops surrender to General George Washington after the battle of Trenton. These Hessians were typical of the highly trained automatons who fought Europe's wars in the eighteenth century. (Copy of 1850 lithograph from the George Washington Bicentennial Commission, National Archives)

"Gentlemen of France, Fire the First Volley"

THE BRITISH were coming. Muskets on their shoulders, fixed bayonets glittering in the May sunshine, rank after rank of red-coated soldiers were advancing at the stately gait of eighty paces a minute, keeping time with their drummers. Artillery fire from the redoubts to the south and north tore gaps in their ranks, but the troops automatically closed the gaps and kept coming. The French troops, standing in line, some behind breastworks, some in the open, waited with loaded muskets.

It was May 11, 1745, the fifth year of the War of Austrian Succession, an extremely complicated affair that had embroiled most of the nations of Europe. The Bourbons of France were again fighting the Hapsburgs, the Austrian Hapsburgs this time. Their troops were invading the Austrian Netherlands, which had been the Spanish Netherlands before Spain lost them in the War of the Spanish Succession. Hermann Maurice, comte de Saxe, the natural son of Augustus the Strong (a mistranslation of Augustus the Potent, a nickname the king earned for the number of his illegitimate offspring), king of Poland, was commanding the French. Saxe had been besieging Tournai when he heard that an army of British, Austrians, and Dutch was approaching. He left 17,000 men to continue the siege and established a defensive line five miles southeast of Tournai. In the center of the line was the village of Fontenoy. The right of the French line was around the town of Antoing with its flank protected by the Scheldt River. The left was behind a forest called the Wood of Barry. Saxe's troops hastily built breastworks in front of Antoing, three earthen redoubts to the left of them and some more breastworks in front of Fontenoy. Another redoubt was built at the edge of the forest in an area where it could not be seen from a distance. He stationed sharpshooters in the forest and put his Irish Brigade, his best infantry, behind the woods. He had 53,000 men and 70 cannons.

Commanding the approaching army was William Augustus, duke of Cumberland, the natural son of King George II of Britain. Shortly after this, Cumberland would command the forces that put down the Scottish revolt under Prince Charles Edward Stuart (Bonnie Prince Charlie) and acquire the Scottish nickname "Stinking Willie." Cumberland also had 53,000 men, but he had 80 cannons. The Dutch and Austrians attacked Saxe's right, but the fire of infantry behind the breastworks and the cannons in the redoubts turned them back. Cumberland sent his cavalry against the unfortified center of the French line, but artillery crossfire from the redoubts cut down the troopers and stopped the charge cold.

Cumberland then formed his infantry into a column of many ranks and marched directly at the French line. On and on they came, while cannons from redoubts on either side blasted them. The French held their fire. Their smoothbore muskets were most effective at close quarters. When the British were a few yards away, according to a story that may be true, an English officer stepped out in front of the column, bowed, and said, "Gentlemen of France, fire the first volley." At that distance, the French would have no time to reload before the English were on them. The French declined the invitation. That was a mistake. The British fired a volley and devastated the French first line, then drove right into the French center. The French second line fired, and Saxe's two elite units, the Swiss Guard and the Irish Brigade, struck each flank of the British formation. After a good deal of gory fighting, Cumberland had to retreat. Tournai fell on June 19, and the French went on to conquer almost all of the Spanish Netherlands.

Armies of Automatons

The almost incredible performance of the British troops, marching with apparent indifference through enfilading artillery that, as it was firing on their flanks, could take out a dozen or more men with a single shot, then walking up to within speaking distance of their enemies, firing a volley, reloading, and continuing their march, was standard in eighteenth-century warfare. It was the result of a reaction to the horrors of the Thirty Years' War and other wars of religion.

The eighteenth century was the age of the Enlightenment, the age of Isaac Newton, who revolutionized physics, and the age of Hugo Grotius, whose *Rights of War and Peace* was a pioneering work on international law. It was a time when philosophers were discussing the rights of man and when everyone was concerned with avoiding excess. The Thirty Years' War had strongly demonstrated the evils of excess. Monarchs like Frederick the Great and Catherine the Great cared little about the rights of man, although they patronized the philosophers. They did, however, care about the prosperity of their states. Ruined countries, like most of the states of Germany after the Thirty Years' War, could produce little tax revenue. Wars should be fought in a way that disturbed civilian life as little as possible. Huge armies should be avoided. They took away men who otherwise would be increasing the wealth of the nation with their work in agriculture and industry. Armies should be small but efficient. Maurice of Orange, Gustavus Adolphus, and Albrecht von Wallenstein had shown how drill makes armies efficient.

One factor that aided this restraint was that wars no longer had the emotional content of the old wars of religion or the future wars of nationalism. Kings now fought to secure more defensible boundaries, for possession of colonies, or to discomfit a rival dynasty. War was in fact the "sport of kings." Ordinary people had little connection with their monarchs' wars. The pawns the monarchs played their games with were professional soldiers—a minority despised by all classes of society.

All the kings of Europe recruited men who were poor and unemployed, men who were not contributing to the national wealth and were not expected to do so. Their recruiters also went after drunkards and petty criminals who were detracting from the common good. The armies signed these men up for long enlistments and drilled them incessantly. They inflicted savage punishments for relatively minor offenses. The object, following a maxim of Frederick the Great, was to make the men more afraid of their officers than of the enemy. Desertions were common, and every effort was made to prevent them. Barracks were invented at this time, because they made it more difficult to desert.[1] Pursuit of a beaten enemy, especially by cavalry, was avoided if possible, because officers felt it was too easy for pursuing troopers to disappear.

Improvements in weaponry at the end of the seventeenth century made possible new tactics. The most important improvement was the bayonet. At first it was merely a dagger the soldier jammed into the muzzle of his musket. But while the bayonet was in place, he could neither load his weapon nor fire it. Later bayonets hung insecurely from a pair of rings that fitted over the barrel of the musket, but finally they were fitted with a socket that locked them to the musket but still allowed the gun to be loaded and fired.[2] With the perfected bayonet, armies were able to get rid of pikes. The soldier's musket was also his pike, and cavalry could no more break through a line of bayonets than it could a line of pikes. Cavalry could always ride around infantry in open country, but the infantry learned to form squares, each side of the square two or three ranks deep. The horsemen could ride around this square of bayonets, but they couldn't do any damage. Special formations and drills to allow pikemen to shelter musketeers from cavalry were no longer needed. And since every soldier had a gun, firepower was greatly increased. This was most important. Use of the bayonet gradually declined to where Napoléon Bonaparte's surgeon general said that for every bayonet wound he treated, there were a hundred caused by guns.[3]

The musket had been vastly improved with the adoption of the flintlock. Handling a smouldering match while loading gunpowder into a matchlock was extremely dangerous. It was also slow: a soldier loading a matchlock had to detach the match from the match holder on his gun, hold the match with his left hand, which also held the gun, and load with his right hand.[4] The flintlock was safer, more reliable, and faster. American soldiers in the Revolutionary War, who were by no means drilled as intensively as European troops, were expected to be able to get off fifteen shots in three and three-quarter minutes.[5] Finally, it was now possible to make night attacks—something the glowing matches of the older guns made impossible. While the infantrymen got these improvements, their colleagues in the artillery got more manageable field pieces. The new guns were almost as light as Gustavus's four-pounders and much more powerful.

To use the new weapons most effectively, the infantry faced the enemy in a long line, only two or three ranks deep. They fired volleys on command, usually a battalion at a time. Soldiers stood shoulder to shoulder to concentrate their fire. They did not aim: they fired in the general direction of their enemies, also standing shoulder to shoulder. The muskets were not accurate, and gunmakers took no pains to make them accurate.[6] All firing was at short range. Accuracy was not important, but speed was. Rifles had been made for almost two centuries but armies did not adopt them, although they were far more accurate than smoothbores, because they were too slow to load. This form of warfare—rapidly repeated volleys at short range—was extremely bloody. Such exchanges of fire tore huge gaps in opposing lines.

Because of these tactics, close-order drill had enormous value. Soldiers marched and performed manuals of arms so long that they became virtual automatons. Their officers were satisfied with nothing short of perfection in the drills. Prussian officers used surveying instruments to determine if the line of a body of troops was straight enough.[7] Parades before the king or other officials were frequent, and woe to any soldier who was less than perfect on these occasions. Parades and the close-order drill that they displayed seemed to develop an almost hypnotic hold on the leaders of armies. Close-order drill is still a feature of military training in all armies, although infantry doing what they were created to do seldom march at all today. And they never do it marching in step in neatly ordered ranks. They ride to combat in trucks, armored personnel carriers, and aircraft. Close-order drill undoubtedly helped the infantry of Frederick the Great be effective, but it probably did not help as much as all the practice they got loading and firing their muskets, aided by the Prussian iron ramrod, which did not, like the wooden ramrods of other countries, break at inconvenient times, rendering the weapon useless except for its bayonet.

Some kings seemed to value their soldiers more for parades than for battles. Frederick William of Prussia had his "Potsdam Giants," a regiment of extremely tall men recruited from all over Europe. On one occasion, his agents kidnaped a tall Italian priest while he was saying Mass and hustled him off to Prussia. Frederick William could not bear to think of his 3,000 giants on a battlefield. He was not a philanthropist, he just considered the men his toys, and he didn't want to lose his toys. He was actually a royal sadist who harassed his effeminate son, Frederick, unmercifully. On one occasion, he was going to have the boy shot, but changed his mind only because the royal guards refused to help the king murder his own son. The son, who became Frederick II, nicknamed Frederick the Great, was a friend of Voltaire and a patron of the arts. He was also a great tactician with an unholy love of battle. And in his battles, he sometimes lost almost half of the men under his command. Frederick was a humanist, but not a humanitarian. To him, soldiers were nothing more than replaceable parts. He had plenty of parts. Ninety percent of the Prussian government's budget went to the army, so he was able to hire huge numbers of soldiers for his small country.[8] When he began the War of Austrian Succession, his army of 150,000 was larger than Austria's.[9]

Other monarchs cared more for the lives of their troops, mostly because it took a long time to train them to the proper degree of robotlikeness. European armies in the eighteenth century avoided living off the land by having fortified magazines stocked with supplies where their armies would be operating. Generals sought to maneuver their enemies away from access to their magazines. Some campaigns were almost bloodless. There were many sieges, but all belligerents accepted that there was a point, when a besieging army had achieved certain positions before an assault, where a besieged army could surrender without disgrace. An extreme example of this sort of siege was the siege of Pizzighetone in 1733. According to a participant, a truce was arranged and:

> [a] bridge thrown over the breach afforded a communication between the besiegers and the besieged: tables were spread in every quarter, and the officers entertained one another by turns: within and without, under tents and arbours, there was nothing but balls, entertainments and concerts. All the people of the environs flocked there on foot, on horse back, and in carriages: provisions arrived from every quarter, abundance was seen in a moment, and there was no want of stage doctors and tumblers. It was a charming fair, a delightful rendevous.[10]

Marshal Saxe, the victor at Fontenoy, one of the bloodiest battles of the century, even wrote that a successful general might wage war all his life without resorting to battle.

This age of limited warfare lasted little more than a century. It received its first challenge in the New World. Warfare as practiced by aboriginal Americans had many limits, but they were limits utterly foreign to the Europeans who came to America. The Indians, for example, never fought to annihilate another nation. They rated a warrior not by the number of enemies he had slain, but by the magnitude of the risks he had taken. A man got far more honor for touching an enemy with his hand than for shooting him with an arrow. On the other hand, the European custom of taking prisoners of war was unknown to the Indians. Prisoners were sometimes adopted into the tribe. Much more often, they were tortured to death, and very frequently, they were eaten.

Faced with what they saw as the diabolical cruelty of the Native Americans, whom they considered an inferior species anyway, the Europeans abandoned all limits and fought to annihilate their enemies. There were too few Europeans in America to depend on professional armies: every boy over twelve was a soldier in emergencies. When the English and the French fought their wars in the colonies, each side with Indian allies, the colonial wars were as unlimited as the Indian wars. The American Revolution was more like a standard European war, but it had a new feature. There were no professional soldiers on the American side, though the Americans tried to develop the Continental army into a reasonable facsimile of one. They even hired the retired Prussian captain Friedrich von Steuben to train the new troops.[11] Steuben, like all the other foreign officers who aided the American cause, is rightly celebrated as a hero of the revolution. But the fact remains that the Americans lost as many battles post-Steuben as they did pre-Steuben. The local militia—soldiers of a type unknown in most of Europe for centuries—were the surprise element and in many ways the key to American success. They were seldom able to stand up to regulars. Bunker Hill was an exception. But the militia wiped out British foraging parties, most notably at the Battle of Bennington; they cut British supply lines, as they did in the Saratoga campaign; they reinforced the Continentals, as they did at Saratoga and the Cowpens; and they destroyed Tory raiders, as they did to Patrick Ferguson's command at Kings Mountain. General Charles Cornwallis had to hole up at Yorktown because the militia prevented him from living off the land. The American Revolution was the first violent incident in a revolutionary age—an age that would change the face of Europe and the conduct of war.

Signs of Change

The times were revolutionary in many ways. James Watt was working on his steam engine; engineers were building networks of roads and canals all over Europe; governments had begun taking censuses of their people; philosophers were proclaiming the rights of man, economists were producing tomes on "the invisible hand" of the free market, and military planners were improving old weapons and devising new procedures for command.

Jean Maritz, a Swiss gun founder in France, made a major improvement in artillery. It had been impossible to cast a gun in which the bore was absolutely centered in the gun, because the mass of molten metal always displaced the core of the mold. Furthermore, precision casting of iron was not possible at this time, so the bore always had to be wider than the shot it was to shoot. Maritz solved the problem by inventing a machine that could drill a precise bore in cast iron. He then cast the guns solid and drilled the bore afterward. So the French got guns that were accurate and, because they fitted the shot more exactly, more powerful than the same-sized guns made by the old method. Guns of the same power as older guns could be made lighter and more mobile.

Then Lieutenant General Jean Baptiste Vacquette de Gribeauval took command of the French artillery. He introduced a screw for more accurately adjusting the elevation of a gun and a new sight that made aiming far more accurate. Gustavus had used cartridges containing the powder needed for one cannon discharge, eliminating the need for the gunner to try to measure the necessary amount of powder in the field. Gribeauval introduced a cartridge containing both powder and shot. Under Gribeauval's direction, the French artillery began using different kinds of projectiles—solid cannon balls, explosive shells, grape shot, and cannister shot—depending on the target and the range. He took the transportation of guns away from civilian contractors and had army gunners haul their guns. He drilled them in unlimbering guns, aiming them, firing them, manhandling them, and hitching them back on the caissons. The drill had the same effect as

Patrick Ferguson

The American Revolution was a strange mixture of the restraint of eighteenth century European dynastic wars and the brutality of civil and colonial wars. Both of these qualities were embodied in one man, Major Patrick Ferguson of the British army. Ferguson had two more qualities seldom found in British officers. He was mechanically gifted and a superb marksman.

While a captain, Ferguson invented a breech loading rifle at a time when the military refused to consider anything but smoothbore muzzle loaders for the common infantryman, and would continue to do so for another half century. Turning the trigger guard on Ferguson's rifle one complete turn unscrewed a plug in the breech. The shooter then inserted a musket ball and gunpowder, screwed the plug back up and fired. Ferguson demonstrated his invention before military authorities in 1776. A contemporary reported:

> Notwithstanding a heavy rain and the high wind, he fired during the space of four or five minutes at a rate of four shots per minute, and also fired (while advancing at the rate of four miles an hour) four times in a minute. He then poured a bottle of water into the pan and barrel of the piece when loaded so as to wet every grain of the powder, and in less than half a minute he fired with it as well as ever without extracting the ball. Lastly, he hit the bull's eye lying on his back on the ground, incredible as it may seem to many, considering the variations of the wind and the wetness of the weather. He only missed the target three times during the whole course of the experiments.

Ferguson commanded a light infantry company and armed his men with the new rifle. But he was wounded at the Battle of Brandywine. While he was in the hospital, the authorities took his rifles out of service. After he returned to duty, Ferguson was promoted and given command of a Tory group in the South, where his ruthlessness earned him a reputation almost as bad as that arch villain, Banestre Tarleton. A Whig militia surrounded Ferguson and his Tories at King's Mountain. Repeatedly wounded, this champion of the rifle led bayonet charge after bayonet charge against the frontier riflemen until he dropped dead.

His actions on the southwest frontier, where he burned out farmers whom he believed were supporting the Patriot cause, contrasted strongly with what happened at Brandywine just before he had been knocked out of action. He was acting as a sniper far ahead of British lines when he saw a tall, distinguished-looking man in the uniform of a high-ranking American officer. Ferguson aimed at the enemy officer, then lowered his rifle and looked for another target. He decided he could not kill such a noble-looking man from ambush.

The American officer was George Washington.

the infantry's musket drill. The French artillery became the best in Europe. Finally, Gribeauval standardized the army's guns. He eliminated many odd-sized guns and kept only a few of the most useful.[12]

The improvements in methods of command were even more important. For years, generals could seldom command more than 50,000 men in a battle. Even standing shoulder to shoulder in the thin lines used in the eighteenth century, the ends of the line would be almost out of sight. Even worse, there was no way a general could signal distant units to make an immediate move. Travel was dangerous and difficult with a large force. An enemy might ambush the end of a long line of marchers and wipe it out before the general knew there was trouble.

The first innovation was the production of accurate local maps. It wasn't until the middle of the eighteenth century that armies had the use of such maps. They were especially helpful because of another innovation: the division. Each division was a little army. It had its own infantry, artillery, and cavalry, as well as such support services as engineers, medical personnel, and quartermasters. Several divisions could be combined to make an army corps—a somewhat larger small army. Staff specialists in the army's general headquarters prepared written orders to the division and corps commanders, who would then use their discretion on how to follow them. In Napoléon's army, orders were sent to the local commanders every night. They, at the same time, would send reports on developments in their areas. Operations areas could be widely scattered. Sometimes dozens of miles separated the units. This allowed a general to confuse his opponents as to his true objective. He could then rapidly concentrate his corps or divisions at a key point.

All of these innovations had been developed by officers of the ancien régime in France. In a few years, they would be wielded by a very different regime, and they would rock Europe for more than two decades. They would also permanently change the concept of war.

10

Not Armies, but Nations

Bringing up a field piece in the Napoleonic Wars. Extensive use of artillery was a hallmark of Napoléon's tactics.

A Contagion from the New World

IN 1709, Louis XIV, the proudest king in Europe, faced his darkest hour. The coldest winter in a century had killed most of the crops and much of the livestock, producing a famine in France. His armies had been defeated on many fronts by a combination of nations that wanted to prevent his grandson from becoming king of Spain. He had sued for peace, but the allies made the terms increasingly harsh as negotiations went on. Finally, they wanted him to help them oust his grandson from the Spanish throne.

"If I have to have war, then I would rather fight my enemies than my children," Louis said.[1]

His enemies, the Austrians, the Dutch, the English, and most of the German states, had overwhelming numbers and were led by the greatest general of the time: the Duke of Marlborough. Marlborough was assisted by Prince Eugene of Savoy, an officer many considered the second greatest general. Louis's army was smaller, dispirited, and exhausted. Most of his generals had a record of failure. In his desperation, the Sun King, the most absolute of absolute monarchs, turned to the only possible source of help:

His people.

In a broadside addressed directly to the population, Louis said, "I have come to ask . . . your aid in this encounter that involves your safety. By the efforts that we shall make together, our foes will understand that we are not to be put upon."[2]

The people responded. Money poured into the treasury and volunteers swarmed into the army. In the showdown at Malplaquet, the still-outnumbered French under Marshal Claude-Louis-Hector Villars gave Marlborough a most Pyrrhic victory. The allies lost almost three times more men than the French. The losses were so high that Marlborough was removed from command. Villars then routed the allies under Prince Eugene, and Louis gained an honorable peace with his grandson still the king of Spain.

The French monarchs soon forgot this demonstration of the power of the people. The people did not, however. When the English colonies in America revolted, the people of France watched sympathetically. So did the monarchy, because Britain was a rival for control of the seas. When the Americans made Lieutenant General John Burgoyne surrender his whole army in the wilds of upstate New York, the king of France made an alliance with the republicans of America. French troops joined the Americans in their country, and the French navy completed the blockade of General Charles Cornwallis's army at Yorktown, which was being besieged by French and American troops.

The young Frenchmen who had served in America returned home convinced that the people of a nation could govern themselves. They returned to a nation where the Estates General had not met since 1614 and where increased taxes would be needed to pay interest on the loans floated to finance the American war. And those taxes would be paid by the common people, not the nobles or the clergy. And all of the French knew that the cause of the American Revolution was taxation without representation.

The king had to turn to the Estates General. When they met, the Third Estate, the commoners, who would have to pay any new taxes, was not in a compliant mood. The commoners refused to meet as a separate order and invited the nobles and clergy to join them. Only a few did. The Third Estate then declared that it was the National Assembly. To dampen the fires of dissent among the Third Estate, Louis XVI ordered the clergy and nobles to join them. He also ordered his Swiss and German mercenaries to assemble at Versailles in case of trouble in Paris. This move enraged the people of Paris. Egged on by radical leaders, the Parisian proletariat on July 14, 1789, stormed the Bastille, an old castle used as a prison, and massacred its garrison. Elements of the army joined the mob and brought artillery, greatly aiding in the capture of the old fortress.

The king ordered his soldiers to withdraw from Paris and Versailles, and the National Assembly created its own military force, the National Guard, composed mostly of middle-class Parisians and commanded by the Marquis de Lafayette. The National Guard also included a substantial number of regular army soldiers, including members of the royal guards. Its officers were elected—an idea brought over from the American militia—although Lafayette had a strong influence on who got elected.[3]

On August 26, 1790, the assembly issued a new constitution with a preface entitled "The Rights of Man." Although the statement took its title from the favorite topic of French political philosophers, it took much of its language from the American Declaration of Independence.[4] The king, in his palace at Versailles, hesitated to ratify the constitution, so Lafayette and the National Guard marched to Versailles and brought the king back to Paris, thus bringing to an end the

French Divine Right of Kings. Henceforth, Louis XVI would be a constitutional monarch.

That, it turned out, would not be long. On the night of June 20–21, 1791, Louis and his family slipped out of Paris and headed for the border of the Austrian Netherlands. They were arrested at Varennes and sent back under guard. On September 14, the National Assembly adopted a new constitution and changed its name to the Legislative Assembly. The majority of the Legislative Assembly began agitating for a war against Austria, the homeland of the unpopular Queen Marie Antoinette. The Parisian newspapers took up the agitation, arguing that a war would unite the nation. Even Louis supported a war; he hoped the Austrians could rescue him.

The Legislative Assembly issued a call to arms on July 11, 1792, and a horde of volunteers enlisted. Austria declared war, and Prussia joined it. In another constitutional change, the assembly became the National Convention, the monarchy was abolished, and Louis and Marie Antoinette were imprisoned. Lafayette, shocked, ordered his troops to march on Paris. They did not, and the marquis crossed the border to Luxembourg.

The convention replaced Lafayette with Charles-François Dumouriez. Dumouriez found that he was commanding a sullen rabble that was opposed to all authority, including his own. When a soldier on parade screamed, *"a bas le general!"* Dumouriez drew his sword and challenged the heckler to fight. The heckler remained silent, and nobody moved. They looked at their new general with respect. Even so, the French prospects did not look good. An army of 42,000 Prussians, 5,500 Hessians, and 4,500 French émigrés under the duke of Brunswick, deemed the greatest general in Europe, had crossed the border. On Brunswick's left were 15,000 Austrians.

The fortress of Verdun fell almost immediately to the Prussians. Dumouriez fell back to the Ardennes while another army under François-Christophe Kellermann moved up from the south to join him. The Austrians attacked one of the passes Dumouriez was trying to defend and forced his troops from the pass. Dumouriez did not retreat, however. Instead, he moved south, united with Kellermann, and threatened Brunswick's flank. Brunswick made a wide-turning movement and got between the French and Paris. The French had been thoroughly outmaneuvered, cut off, and could be expected to panic. Brunswick ordered his artillery to bombard the French position. The poet Johann Goethe witnessed the cannonade, "the violence of which at the time it was impossible to describe," that made the battlefield tremble. Brunswick looked through his telescope to see if he could see any fleeing French. He saw nothing. The untrained rabble must have run at the first shots. He ordered his infantry to advance. The French had not fled. Instead, they wheeled up their artillery and opened fire. Before the revolution, the French artillery was the best in the world. It still was. Brunswick halted the advance before it reached musket range. *"Hier schlagen wir nicht"* (We do not fight here) he told his staff.[5]

At this time, the French had 36,000 men and 40 cannons. Brunswick had 34,000 men and 58 cannons. The first test of the revolution had been a contest of wills between revolutionary fervor and Prussian discipline. Fervor won. And the revolution had not yet unleashed its strongest weapon.

The Levee en Masse

On August 23, 1793, the National Convention passed a law that ended the era of unlimited warfare. It decreed that "all Frenchmen are permanently requisitioned for service into the armies. Young men will go forth to battle; married men will forge weapons and transport munitions; women will make tents and clothing and serve in hospitals; children will make lint from old linen; and old men will be brought to the public squares to arouse the courage of the soldiers, while preaching the unity of the Republic and hatred against Kings."[6]

The mobilization of France wasn't quite that complete, of course. But the decree resulted in an enormous army that the forces of reaction couldn't handle. By January 1, 1794, the French army numbered 770,000.[7] With overwhelming numbers, the French revolutionaries drove into the enemy countries. At Handschotten, on September 8,

1793, 50,000 revolutionaries routed 15,000 soldiers of the old regime. The odds were only slightly less overwhelming at Wattignies a month later—45,000 to 18,000.[8]

The prerevolution creation of divisions and corps made it (barely) possible to command these masses of men. French tactics were not like anything seen in western Europe for a long time. The French were said to deploy in columns instead of the traditional line, as if attacking in columns was something new. It wasn't. Sir William Howe organized his troops in columns for his last assault on Breed's Hill (the so-called Battle of Bunker Hill) in the American Revolution. There was much learned discussion of column versus line among military academicians. A column, with a short front and a depth of many ranks, limited the number of muskets that could fire at one time, but it exposed fewer troops to the musket fire of the enemy. It could be a very poor choice if the enemy had cannons behind the point to be attacked. On the other hand, if the column got within bayonet range, it was sure to break through a line only three ranks deep.

Actually, those early French columns were not much like the columns of Howe's redcoats at Bunker Hill. They were more like mobs of armed men crowded together and rushing at the enemy.

The French revolutionaries were also said to make extensive use of light infantry. British light infantry in the American wars, such as the French and Indian War and the American Revolution, were highly trained troops with lightened equipment. They were good marksmen, capable of carrying out intricate maneuvers at a run, and expected to think for themselves. The French light infantry were not highly trained, were indifferent marksmen, and didn't carry out intricate maneuvers at any kind of pace. They did, however, think for themselves. They ran toward the enemy, scattered over the front, and took advantage of any kind of cover, firing from behind trees and buildings. In the language of the times, they deployed as skirmishers.

Napoléon

When Napoléon Bonaparte took command of the French army, he built on these spontaneous revolutionary tactics and on the prerevolutionary organization of divisions and corps. He was blessed with enormous energy and a prodigious memory. He slept from 8 P.M. until 1 A.M., when the reports from his corps and division commanders began to come in. He didn't need a staff to tell him where each unit was and what it faced in the way of enemy troops and terrain features. He had that information in his head. His huge, wide-spread armies were able to menace several points at once. When ready, he could concentrate his troops, often using forced marches and night marches, at the point he chose. His skirmishers spread out to hold the enemy in place while he brought up his artillery and infantry columns. Napoléon began his career as an artillery officer, and he used his cannons more effectively than any general before him. He'd concentrate his cannons where he wanted to break through an enemy's lines and literally blow breaches in their lines. The columns would then charge through.

Napoléon is supposed to have said, "In war, the *moral* is to the physical as three is to one." The English word for *moral* is morale. Napoléon's troops had that because, although they were fighting for a dictator who soon made himself emperor, they believed they were bringing liberty, equality, and fraternity to the oppressed victims of the absolute monarchs. And they convinced many of the people in the lands they invaded. As a result, Napoléon filled his armies with eager volunteers from the conquered countries.

Morale was a factor Napoléon's enemies found hard to match. They did, however, copy the French organization of divisions and corps, as well as the French tactics of skirmishers and columns. They also made an effort to match the French numbers, and this had the most lasting effect on the conduct of war. Continental powers introduced conscription, as the French had. Future wars would no longer be merely conflicts of professional soldiers. They would be struggles between whole nations.

At the eastern and western ends of Europe, a peoples', rather than a soldiers' war had already begun. Scorched earth, weather, and Russian guerrillas devastated Napoléon's Grand Army in 1813, bringing an ignominious end to his invasion of Russia. In Spain, at the same time, Spanish guerrillas and British regulars cooperated in driving the

French off the peninsula. The revolutionary fervor of the French soldiers had ignited a new fervor in the nations of Europe: nationalism.
The revolutionary armies had given the European peoples the morale
their rulers couldn't provide.

After the wars, the European powers followed the Prussian system
of enrolling the drafted men in a series of reserves. If a problem looked
too big for the active duty army, the government could call up
the ready reserves—men who had recently completed their military
training. In a dire emergency, it could call up all trained and able-
bodied men.

But the next big war in the Western world did not take place
in Europe. It happened in America, and it, too, was no mere fight
between professional soldiers. Professional soldiers had a rather small
role in it, and because of that, the American Civil War introduced
many changes in the conduct of war.

11

The American Civil War

The Gatling gun, the first successful mechanical machine gun, was introduced in the American Civil War.

The First Modern War

FIELD MARSHAL HELMUTH VON MOLTKE, chief of the German General Staff, discussing the American Civil War, once said that there was nothing to be learned from "two armed mobs chasing each other around the country."[1] Moltke was no Colonel Blimp. He didn't spend all his life in Prussia. At one time, he was sent to Turkey to help the Turks in their fight against Mehmet Ali, the sultan's rebellious viceroy in Egypt. He had been around; he was a keen student of war, and, as we'll see, he created another turning point in warfare. In one respect, though, Moltke was a typical European officer. He believed that the world outside of Europe was inhabited, if not by utter barbarians, by backward colonials, and nothing that happened there would happen the same way in Europe. Besides, there was no aristocratic officer class in America, and the soldiers in the "two armed mobs" were the rankest sort of amateurs.

The last point was absolutely true. None of the participants in what was to become the United States' bloodiest war had received anything like Prussian military training. Many of the high-ranking officers, men like Ulysses S. Grant and William T. Sherman, had interrupted their military careers with a variety of civilian pursuits. Others, like Philip H. Sheridan, had been young junior officers when the war began. Almost all of the enlisted men were total amateurs who had received the most hasty training before being rushed to the front lines.

Because they were amateurs, men who were not steeped in military traditions and traditional practices, they were able to adapt the many new mechanical marvels of the late Industrial Revolution to warfare.

There had been steam trains before the Civil War. During the Crimean War of 1853–1856, the French had even built a short railroad. But no generals had ever used railroads to move huge numbers of troops hundreds of miles the way Union and Confederate generals did in the Civil War. There had been telegraph lines before, but they had merely run from national capitals to army headquarters. There were no telegraph stations under tents and lean-tos on the battlefields, as there were in the Civil War. Napoléon Bonaparte managed to coordinate the actions of armies separated by dozens of miles. Thanks to the telegraph, Union and Confederate headquarters coordinated armies separated by hundreds of miles. The French Revolution introduced observation balloons, but Napoleon disbanded the balloon corps. Except for the Austrians' unsuccessful attempt to bomb the besiegers of Vienna with balloons, lighter-than-air craft were neglected until the Civil War.

Some innovations in the Civil War were entirely new. There were no practical machine guns in previous wars, nor any repeating rifles. No submarine in a previous war ever sank an enemy ship. In fact, except for David Bushnell's *Turtle* in the American Revolution, there were no earlier submarines. Both the Union and Confederate navies employed several, and the C.S.S. *Hunley* actually sank the U.S.S. *Housatonic*. As we've seen, Yi Sun Shin used ironclad warships in 1592, but there was no conflict *between* ironclads until 1862. The Confederates invented a new class of armored warship, the ram, but the Union built ironclads in a wide variety of types—battleships, monitors, and a swarm of armored river boats—that were instrumental in cutting up the Confederacy. The fate of the wounded in the Civil War was grim, indeed, but it was better than that of those wounded in the previous wars of the past few thousand years. Anesthetics were used for the first time, and Clara Barton provided the first decent nursing care American soldiers had ever had. The enormous demand for equipment of all kinds jump-started mass production in many industries. Ready-made clothing, for example, is a by-product of the need to produce millions of uniforms during the Civil War.[2] Ironically, one U.S. industry that did not derive its switch to mass production from the Civil War was the firearms industry. It had been mass-producing guns years before the war. What the British called the "American system of manufacture" started in Connecticut and Massachusetts around 1820. It was based on machine-made, interchangeable parts and was developed to make up for a shortage of skilled gunsmiths in the young republic. Europeans became aware of it at the London Exhibition of 1851, where Samuel Colt demonstrated the advantage of interchangeable parts. The British switched to the "American system" after that, and other Europeans soon followed.[3]

At least as important as the new hardware of war was the way the amateur warriors of America adapted to its use. It took a while for them to learn to live with the universal use of rifles, but when they did, the world got a preview of World War I. The American soldiers made the traditional cavalry charge obsolete, but Moltke and his many duller colleagues didn't get the word for another half-century. The Prussians had the breech-loading Dreyse rifle a generation earlier, but they didn't take full advantage of it. The Americans, with their Sharps breechloaders (vastly better guns than the Dreyses) and their Henry and Spencer repeaters, did use them to best advantage.

So many new weapons and practices were introduced that the American Civil War is a huge turning point in warfare. It was also, of course, a turning point in the history of the United States and of the world. At the end of the war, the United States had the largest and most powerful army and navy on earth. Its industrial capacity exceeded that of either Britain or Germany. Its power was recognized by France when the United States sent troops to the Mexican border and Napoléon III called his troops home, leaving Maximilian in an impossible situation. It was recognized by Britain, when the United States sided with Venezuela when that South American country had a border dispute with the British. In spite of that, you can still find American authors saying their country was not a great power until after World War I.

Let's take a closer look at some of the many ways the Civil War changed warfare.

The Ironclads

The idea of armored ships disappeared between the death of Yi Sun Shin and the nineteenth century. Europeans in what has been called the Great Age of Sail protected their ships with enormously thick hulls of seasoned oak. That protection was so effective that fighting ships closed with each other to pistol-range before firing their guns. The guns fired cast-iron cannon balls almost exclusively because only the hard, heavy solid shot had a chance of penetrating a ship of the line's hull. They also used bar shot—two halves of a cannon ball connected by an iron bar—and chain shot—similar to bar shot but with a chain in place of the bar. These projectiles were effective in cutting rigging and knocking down masts. Most naval battles were won by disabling the enemy ship and boarding it with pistols, pikes, and cutlasses. In some ways, naval warfare had not changed much from the time of the Romans.

A French artillery officer changed that. In 1822, Colonel Henri Joseph Paixhans published the treatise *Nouvelle Arme*, which called for a new use of an old weapon. Explosive shells had been used since the sixteenth century, but only in land warfare. They were strictly antipersonnel weapons. Explosive shells had neither the weight nor hardness of solid shot, so in sieges, they were fired over the walls of fortresses. Solid shot was used to batter down the walls. In the field, shells were fired at troops too far away to hit with grape shot. Shrapnel, a type of shell filled with lead balls as well as gunpowder, invented by British lieutenant Henry Shrapnel, was especially effective against troop formations. Shells were not used by ships, however. The comparatively light shell would not penetrate a warship's hull, and the range of naval cannonades was so short that grape or cannister shot would be more effective against enemy sailors.

Paixhans, a soldier, did not have his thinking inhibited by naval tradition. He pointed out that a shell didn't have to penetrate a ship's hull to cause damage. If it lodged in the wooden sides and exploded, it would make a much bigger hole than a cannon ball. The explosion would also throw hot metal fragments and blazing wood splinters far and wide. They would wound and kill sailors and start fires. Sails, tarred ropes, and the wooden hulls of ships were all quite inflammable. Furthermore, naval shells could contain incendiary mixtures as well as, or instead of, gunpowder. Among the first sailors to try Paixhans's idea were the Russians. In 1853, in the Battle of Sinope, a Russian squadron using shells burned a Turkish squadron of twelve ships.

After that, all navies became interested in ships that wouldn't burn. The French were the first off the mark. In the Crimean War, they used three floating batteries, armored with sheets of wrought iron, to bombard Russian forts. The Russians used both shot and shell, but the French batteries suffered no significant damage.

At the outbreak of the Civil War, the U.S. Navy was almost entirely composed of wooden-hulled, steam-powered ships. The first steamers were side-wheelers. The paddle wheels posed problems, however. They made it impossible to mount guns along much of the ship's side, they were highly vulnerable to enemy fire, and they made handling the ship difficult when the engines were off and the ship was under sail. In 1841, Captain Robert F. Stockton of the U.S. Navy called on John Ericsson, inventor of the screw propeller, and induced him to design the power plant for a new U.S. warship. Ericsson, a Swede, moved from England to the United States, became a U.S. citizen, and created the U.S.S. *Princeton*, the world's first propeller-driven warship.[4] From that point on, the oceangoing paddle-wheel warship was a thing of the past as far as the U.S. Navy was concerned.

When the Civil War began, the commander of the Gosport, Virginia, Navy Yard, ordered the ships and equipment there destroyed so the Confederacy couldn't use them. The workers, most of them Confederate sympathizers, didn't do a very good job.[5] Among the things the Confederates managed to salvage was the hull of the forty-gun steam frigate *Merrimack*. Confederate naval architects turned *Merrimack* into something entirely new and renamed it the C.S.S. *Virginia*. They gave it a sloping superstructure composed of two, two-inch-thick layers of wrought iron and a one-inch-thick belt of iron around the hull that extended to three feet below the waterline. The superstructure was 178 feet long and 24 feet above the waterline. Partly because of the weight of the iron, the completed vessel sat low in the water and had a draft of twenty-two feet. It could not operate in shallow water, and its low freeboard would make ocean travel hazardous. In action, an observer said it looked like a barn roof floating in the water.[6] For armament, *Virginia* had eight smoothbore cannons, four on each side, and two seven-inch rifled cannons on the bow and the stern. The rifled guns fired elongated projectiles. If the projectiles were shells, they carried more gunpowder than spherical shells, and if they were solid shot, they were heavier than cannon balls and had more penetration. *Virginia* also boasted a new version of the oldest naval weapon: a metal ram attached to its bow. *Virginia* was the first of a new class of warships, called rams, that were used exclusively by the Confederacy. Some European navies later built rams, but they were built quite differently, with most or all of their guns firing directly over the bow, like the Renaissance galleys.

Reports of the construction of a Confederate ironclad caused a near-panic in Washington. Navy Secretary Gideon Welles knew about the ironclad French batteries, but he hadn't wanted to experiment with such newfangled notions. News of the Confederate ironclad changed that. Congress ordered the construction of three ironclad vessels. Two, *Galena* and *New Ironsides*, looked like conventional ships, although *New Ironsides* mounted on a swivel the heaviest gun ever put to sea up to that time. The third ship was a revolutionary design by Ericsson, the U.S.S. *Monitor*.

On March 8, 1862, the C.S.S. *Virginia* steamed into Hampton Roads and confronted five Federal blockading ships: *Minnesota, Roanoke, St. Lawrence, Congress*, and *Cumberland*. The last three were sailing ships, among the last in the U.S. Navy. The heavy, underpowered Confederate ram chugged toward *Cumberland*, firing as it advanced. *Cumberland* fired back, but its cannon balls merely bounced off the iron monster and its shells exploded harmlessly. *Virginia* hit the sailing ship with its ram and tore a seven-foot hole in its side. The old frigate went to the bottom. *Congress*, meanwhile, fired broadside after broadside at the ironclad, "having," according to a Union observer, "no more effect than peas from a popgun."[7] After finishing off *Cumberland*, *Virginia* turned on *Congress*. It had to rely on broadsides as the iron ram had been wrenched off when it backed away from *Cumberland*. One shot hit the Union ship's powder magazine and blew it up.

The Confederate ironclad was not entirely unscathed in this first battle. Two of its guns had been knocked out, two of its gun crews killed, and several more wounded. One of the wounded was its skipper, Captain Franklin Buchanan. Every fitting on its deck had been shot away, as well as part of its funnel, filling the ship with smoke. But although *Virginia* had been hit ninety-eight times, not one shot penetrated its armor. *Minnesota*, the steam-powered flagship of this Union squadron, had run aground trying to help the other ships. *Virginia* tried to finish that enemy off, but its deep draft prevented it from closing with the stranded Yankee ship. The Confederates decided to finish the job the following day.

It was a good day's work. Two major federal ships had been destroyed and a steam-powered battleship had been run aground. Union officers sent frantic telegrams to Washington. Replies told them that the U.S.S. *Monitor* was en route to help them. This was small consolation, because *Monitor* was a completely unknown quantity.

The next morning, *Virginia* returned and headed for *Minnesota*. Confederate sailors saw a strange-looking craft beside *Minnesota*. As a Confederate midshipman said later, "We thought at first it was a raft on which one of the *Minnesota*'s boilers was being taken to shore for repairs."[8] Then an eleven-inch smoothbore poked out of the "boiler" and fired. The "boiler," which someone else said looked like a cheese box, spun around, and another gun poked out and fired. *Monitor* was a perfect example of Ericsson's eccentric genius. When he designed *Princeton*, the first propeller-driven warship, he gave it a propeller based on the Archimedes' screw. That type of propeller was later junked and a more efficient wheel type (the kind still used) was substituted for it. The engines driving *Princeton* were unlike anything ever seen before or since. Instead of pistons, it had rectangular plates that swung back and forth "like a barn door on hinges."[9] *Monitor* had the world's first rotating gun turret. It mounted two 11-inch smoothbores firing 175-pound cannon balls. The guns fired alternately. A heavy, crank-shaped bar blocked a gunport while the gun was being loaded.[10] There was only one problem. Ericsson had not provided a way to stop the turret at a precise spot. *Monitor*'s gunners not only had to hit a moving target, but they also had to hit it with a moving gun.[11]

But hit it they did. *Monitor*'s big smoothbores, considerably less advanced than *Virginia*'s rifled swivel guns, cracked the Confederate ship's armor and popped the rivets out of its funnel, again filling the gun deck with choking smoke. The Confederate ship managed to do no more damage to its opponent than occasionally dent the turret's four-inch-thick armor. Once, during the two-hour slugging match between the ironclads, *Virginia* ran aground. *Monitor*, lighter, faster, and of shallower draft, closed in for the kill, but the big Confederate ram broke loose and steamed into deeper water. The fight went on. At one point, *Virginia* tried to ram *Monitor*, but the Confederates' ram was missing, and all the ramming did was to start some leaks in *Virginia*'s bow. A shell from *Virginia* burst on the Yankee ship's pilot house—a tiny, boxlike structure on its deck, while the captain was observing the fight. He was wounded, and *Monitor* stopped firing for a short period. *Virginia* took advantage of the pause to steam back to Norfolk for the protection of the Confederate forts.

The Confederates claimed a victory because the Yankees had stopped firing, and the Yankees claimed a victory because *Virginia* retreated. Tactically, however, the first fight between ironclads was a draw. Strategically, it was a Northern victory, because *Virginia* never again menaced Union ships. When Confederate troops were driven out of Norfolk in May 1862, they blew up *Virginia*.

The Confederates built more rams, including the C.S.S. *Tennessee*, the biggest of them all. During the Battle of Mobile Bay, the U.S.S. *Chickasaw*, one of some thirty *Monitor*-class ships the Union built, hung on *Tennessee*'s rear and sent shot after shot into the same place on its armor. The last shots penetrated its armor and wrecked its engines. *Tennessee*'s commander, the same Franklin Buchanan who had commanded *Virginia* on its first day of battle, now an admiral, had to strike his colors.

Mobile Bay, incidentally, was the first large-scale use of what then were called "torpedoes" and today are called marine mines. There were two types: those detonated electrically from shore—an invention of Samuel Colt of revolver fame—and those by contact with a ship. One of the monitors leading Admiral David Farragut's squadron struck a mine and sank immediately. Farragut's reaction gave the U.S. Navy a slogan it has treasured ever since: "Damn the torpedoes! Full speed ahead!" (Modern torpedoes, invented in 1866 by a Scotsman and an Austrian, were called "locomotive torpedoes." They are discussed in more detail in Chapter 14.)

Moltke may have thought there was nothing to be learned from the American armies during the Civil War, but the British had no such illusions. *The Times* of London, after pointing out that Britain had two ironclads, said, "There is not now a ship in the English navy apart from these two that it would not be madness to trust to an engagement with that little *Monitor*."[12]

Before the war ended, both the Union and Confederate navies had far more ironclads than all the non-American navies combined: twenty-one Confederate and fifty-eight Union ironclads.[13]

Rifles, Little and Big

The rifle was not a new weapon. Rifles had first appeared in the fifteenth century, and they were in many respects greatly superior to smoothbore muskets. The interior of a rifle's barrel is lined with spiral grooves. These grooves spin the bullet as it moves down the barrel. The spinning bullet has gyroscopic stability, and the spinning motion also prevents an unbalanced buildup of air density in front of the projectile. This uneven buildup ruined the accuracy of the smoothbore musket. According to Major George Hanger of the British army, who fought in the Revolutionary War, "A soldier's musket, if not exceedingly ill-bored (as many of them are), will strike the figure of a man at eighty yards; it may even at 100; but a soldier must be very unfortunate indeed who shall be wounded by a common musket at 150 yards, provided his antagonist aims at him."[14] A rifle had an accurate range approximately ten times that of a smoothbore musket. In the Revolutionary War, Americans used riflemen as light infantry and skirmishers, but the overwhelming majority of American soldiers had smoothbore muskets.

Few rifles were used by any army before the Crimean War. The rifle didn't really come into its own until the Civil War. Rifles were too slow to load. There is a legend, beloved by American gun enthusiasts, that European riflemen had to hammer their lead bullets all the way down the barrels of their rifles, because the bullet had to conform to the gun's rifling. On the other hand, according to the legend, American riflemen in the eighteenth century wrapped their bullets in a greased patch that "took" the rifling, giving the bullet its spin, but was much quicker to load than a plain lead ball. The truth is that European riflemen had been using greased patches before many Americans even had rifles. And even using the greased patch, the rifle was slow to load. One reason is that black powder, or gunpowder, the only known explosive at the time, leaves a heavy black residue in a gun's barrel. Unless a muzzle-loading gun is cleaned frequently, it becomes almost impossible to load with an ordinary bullet. A British officer, a British gun maker, and a French officer all came up with similar ideas for a new type of bullet that would make possible a muzzle-loading rifle as easy to load as a smoothbore. Apparently, the British officer, Captain John Norton, had the idea first in 1823; the gun maker, William W. Greener, developed an egg-shaped expanding bullet in

1836, but its accuracy was poor. In 1849, the Frenchman, Captain Charles Claude Etienne Minié invented the expanding bullet that was adopted all over the world.

The idea for what American troops called the "minnie ball" was a bullet that was smaller than the diameter of the bore, so it would slide down the barrel easily. The bullet was somewhat conical, rather than spherical, and it had a hollow base. Minié's version had an iron cup in the base. When the powder exploded, the blast would force the cup into the bullet, expanding the projectile into the rifling. During the Civil War, it was discovered that the iron cup was unnecessary: the lead bullet would expand without it. The conical shape of the bullet gave it more surface to bear against the rifling and more efficiency in cutting through the atmosphere. In other words, both range and accuracy were improved. Smoothbore muskets rapidly disappeared during the Civil War.

As a result, tactics that had been standard from the time of Napoléon and Frederick the Great, even Gustavus Adolphus, became suicidal. First to go was the practice of massing artillery on the front line and blasting gaps in the line of infantry. This tactic had been practical because the effective range of cannons was so much greater than that of smoothbore muskets. However, smoothbore cannons had no such advantage over rifles, so the cannons had to move back or fire from behind earthworks.

Infantry also adopted earthworks. Pictures of Civil War battlefields at the time of the war are eerily reminiscent of the trench networks of World War I and the trench-and-bunker fortifications of the Korean War. Before the end of the war, all the fighting men were sheltering below the surface of the earth.

It took too many months and too much blood to reach that stage, though. Through much of the war, infantry attacks were conducted about the same way George Washington's troops performed. Long lines of men with rifles on their shoulders marched up to the enemy positions, occasionally stopping to fire a volley. But when this happened in the Civil War, the enemy was not waiting to see the whites of their attackers' eyes. They began shooting at the attackers when they were half a mile away. And they were hitting what they shot at. In 1860, the population of the United States was around 32 million— a bit more than a tenth of what it is today. Out of that population, 200,000 men were killed in battle, and another 400,000 died of disease or hardship.[15] Or, to put it another way, the enormously smaller United States of the Civil War, saw just about as many American soldiers die as those who died in World War I, World War II, the Korean War, and the Vietnam War *combined*. To a very large extent, that was the result of the universal use of rifles.

The range and accuracy of muzzle-loading rifles vastly increased the infantry's firepower. But the Civil War also made the muzzle-loading rifle obsolete. The war introduced two new types of small arms that multiplied the muzzle-loader's firepower: the breech-loading single-shot rifle and the repeating rifle. To load a muzzle-loader, it is not necessary to stand, as has often been said, but it is awkward and a lot slower to load it without standing. A soldier using either a single-shot breechloader or a repeater could operate his weapon perfectly from the lowest of prone positions.

The premier breechloader in the Civil War was the Sharps, which later became the favorite arm of western buffalo hunters and eastern target shooters. Colonel Hiram Berdan raised a regiment of Sharps-armed riflemen that became known as Berdan's Sharpshooters. They were excellent marksmen, but the "sharp" part of the Sharpshooter's name derived from the rifle's inventor, Christian Sharps, rather than from the precision of the regiment's shooting. The original Sharps used paper cartridges, but its breech block fit so tightly there was little gas escape at the breech. Gas leakage was the reason most other breechloaders failed before the invention of metallic cartridges. That tight breech block and the strength of the action allowed the Sharps rifle to handle heavy powder charges. It was an extremely powerful, long-range weapon that became even better when metallic cartridges were introduced.

The Civil War repeating rifles and carbines, the Henry and the Spencer, were less powerful than the Sharps but could be fired even faster. After firing a Henry, the shooter swung a lever that was part of the trigger guard down and back and fired again. The Spencer required swinging down a lever but also pulling back a hammer to fire another shot; it could also be reloaded with a seven-shot magazine.

At the Battle of Chickamauga, Colonel John Wilder's "Lightning Brigade" of mounted infantry was holding a bridge against Nathan B. Forrest's cavalry division. Wilder's men had Spencers. A witness said that when the Confederate column charged the bridge the fire of the Spencers sounded like a drum roll and the head of the Confederate column seemed to disappear. Forrest's men continued to rush forward, but when they reached a certain point, they appeared to sink into the earth. Not until Forrest's troopers brought up artillery were they, with the help of two Confederate infantry divisions, able to cross the bridge.[16]

In this battle, Forrest's division fought dismounted. This was how all cavalry, Union and Confederate, fought in the Civil War whenever they got into a major battle. The range and accuracy of the rifle made the old-fashioned cavalry charge suicidal. John Singleton Mosby, a Confederate guerrilla, always fought mounted, but he didn't fight pitched battles. He specialized in small actions—raids, ambushes, and surprise attacks. Mosby's men used repeating carbines, but their favorite weapons were revolvers. Each man carried at least two of them with extra loaded cylinders for quick reloads. They seldom used the saber, which Europeans considered the basic cavalry weapon.

The rifles men carried were not the only rifles in the Civil War. There were also the rifles horses towed—rifled artillery. The rifled field guns had greater range and accuracy than the smoothbores. Their elongated projectiles—possible because of the stability provided by the rifling-imposed spin—were heavier, caliber for caliber, than round projectiles of the smoothbores. They had more penetration and their shells had more explosive power. The range of infantry rifles forced the cannons to move back behind the front lines. Rifling let the cannons move back and still be effective.

Using artillery at long range led to another innovation: aerial observation. Union armies sent observers up in balloons to see where their shells were striking and signal what corrections should be made. Aerial observation was not important with short-range Napoleonic artillery tactics, but from the Civil War on, it became increasingly necessary.

Machine Guns

Machine guns, like aerial observation, made a small beginning in the Civil War, and like aerial observation, they would become increasingly important. At one point, in fact, they would utterly dominate the battlefield. When the Civil War began, a number of inventors came forward with modernizations of one of the earliest types of firearms: the volley gun. Volley guns were large collections of muzzle-loading barrels mounted on some kind of cart. Sometimes, the barrels were arranged so the weapon could fire several volleys. The downside was that the more shots a volley gun could fire, the longer it took to reload it. For this reason, these weapons were never as satisfactory as a cannon firing grape shot.

A Civil War modification of the volley gun was the Billinghurst Requa (BR) battery gun. The BR gun had twenty-five rifle barrels mounted in a row. The great improvement over the volley gun was that, besides the rifling, it was a breechloader. Instead of the paper cartridge used in muzzle-loading rifles, it had steel chambers loaded with powder and bullets. These fitted on a breech mechanism and all twenty-five barrels could be loaded with one motion. A train of powder ran to the touch holes on each chamber. One percussion cap fired the powder train, and the battery gun fired a ragged volley of twenty-five shots. With a number of loaded breech mechanisms, the BR gun could fire seven volleys a minute.

Detachable steel chambers, like those on the BR gun, were used in a number of early machine guns. The Agar "coffee mill" gun used such chambers with percussion caps attached. Loaded chambers were dropped into a hopper while a gunner turned a crank on the side of the gun. The crank powered machinery that chambered a round, fired it, ejected the empty chamber, and loaded a new chamber. It didn't see much use, however, because the Union military authorities distrusted such newfangled ideas. The Confederates, being the weaker side, were more receptive to nontraditional weapons. One was the Williams machine gun, which fired one-pound shells at the rate of sixty-five shots a minute. Picket's Division, which made the ill-fated charge at Gettysburg, had a battery of Williams guns. The Williams gun probably saw more action than any other hand-cranked gun in the Civil War. The future, at least the near future, however, belonged to the brainchild of a North Carolina inventor who had moved to the North.

Richard Jordan Gatling was a physician, but he never practiced medicine. He was a professional inventor. He began working on a machine gun in 1851 and had the design far enough along to patent in 1862. At first, he used paper cartridges, but they did not stand up to the stress of being loaded mechanically. Gatling switched to steel chambers and then to the newly developed metallic cartridges. His first gun had six barrels and ammunition was fed into it by gravity from a magazine mounted above the gun. Turning a crank revolved the barrels. A cartridge entered behind the top barrel. As the barrels revolved, the cartridge was chambered, fired, and ejected. By using multiple barrels (some later models had ten), Gatling avoided a major problem of early machine guns: overheating. Using six or ten barrels allowed the other five or nine to cool while one was firing. Not that they had a lot of time to cool. With a strong gunner turning the crank, Gatling's fully developed gun could fire at a rate of 1,000 rounds a minute.

Gatling offered the gun to the federal government, but because of his Southern birth, the authorities suspected that the gun was some sort of Confederate trick. Major General Benjamin Butler, a Massachusetts lawyer and "political general," had no military training at all, but he didn't lack common sense. He bought twelve of the guns from Gatling, using $12,000 of his own money and used them effectively during the siege of Petersburg.

After the war, the United States adopted the Gatling gun, along with Britain, Russia, and several other countries. France bought a few, too, although the French relied on their "secret weapon," the Montigny mitrailleuse, an awkward volley gun that they mishandled during the Franco-Prussian War. The Gatling gun gradually faded away during the early twentieth century in competition with automatic machine guns like the Maxim and the Browning, but it was revived after World War II in the United States. Its latest incarnations are the 20 mm Vulcan and the .30 caliber Minigun, which can fire at the astounding rate of 6,000 rounds a minute.

Railroads

Railroads had been around for a long time, but the American Civil War was the first time they had been used extensively for troop movements. One reason for this, of course, is that except for Russia, no European country had the kind of distances to be found in the United States. In 1860, the United States had 30,000 miles of railroad—longer than the railroad mileage of all the rest of the world combined.[17] Because there were railroads, Civil War generals were able to move troops in ways that would have boggled the minds of commanders in earlier wars. In 1863, for instance, Braxton Bragg, retreating from the army of William Rosecrans, planned to lure the Yankee general into a trap in the hills of northwest Georgia. Outnumbered at the beginning of the campaign, he asked for reinforcements from all over the South. From Mississippi came refugees from the fall of Vicksburg, and from Knoxville, Tennessee, he got troops who had recently been harassing the Yankees under Ambrose Burnside. Robert E. Lee sent the able James Longstreet and his whole corps from the Army of Northern Virginia. The trip from Virginia to the backwoods of Georgia took the whole corps no more than three days—a trip of 300 miles as the crow flies, and the railroads were much less direct than any crows. In a matter of days, Bragg had assembled an army of 70,000 to meet Rosecrans's once numerically superior army of 57,000.[18] Napoléon would never have thought that possible.

But railroad strategy was in its infancy. In a short time, Moltke would add something brand new.

12

War by the Timetable

Krupp's 75 mm field gun. Prussian artillery thoroughly outclassed the French guns in the Franco-Prussian War.

"Your Majesty, It Cannot Be Done"

IN THE SUMMER OF 1914, Europe had achieved the ultimate balance of power. All the continent's great powers were tied to each other. France and Russia had a treaty requiring mutual support in time of war. France and Britain had another agreement, the *entente cordial*, for mutual support. Austria-Hungary and Germany were similarly united, with Austria-Hungary a somewhat junior partner. Russia had always seen itself as the protector of all small Slavic nations, especially those that were predominately Orthodox by religion. There was so much mutual support that observers had long said Europe was a powder keg that needed only one small spark to set off a world-shaking explosion.

141

The spark was supplied by a .32 caliber pistol in the hands of an eighteen-year-old assassin in Sarajevo. The assassin was Gavrilo Princip, an ethnic Serbian who lived in Bosnia-Herzegovina, a province that had been annexed to Austria-Hungary. The victims were Archduke Franz Ferdinand, heir to the throne of Austria-Hungary, and his wife, Countess Sophie. The Austrians loudly accused the Serbian government of ordering the assassination. At this point, they had no proof, but they happened to be right. Princip and seven other young men had been recruited by agents of Dragutin Dimitrievíc, head of the Serbian military intelligence, known—and feared—as Apis, his code name. Dimitrievíc was Europe's most experienced regicide. Strangely, the reason Apis wanted to kill the archduke was that Franz Ferdinand was known as a champion of the empire's Slavs. (His beloved Sophie was a Czech.) If Franz Ferdinand became emperor, the Serbs might not be able to convince the South Slavs that they should leave Austria-Hungary and join Serbia.[1]

Regicide was, of course, abhorrent to all the crowned heads of Europe, but especially to Wilhelm II, kaiser of Germany, who gloried in his royal status. The kaiser told the government of Austria-Hungary he would back any action it took against Serbia. Austria-Hungary sent Serbia an ultimatum it could not have agreed to without becoming a protectorate of Austria-Hungary. Russia, seeing a threat to one of its client states, issued a mobilization order. Russian troops began massing on both the Austro-Hungarian and the German borders, because Germany had that treaty with Austria-Hungary. Germany, as well as Austria-Hungary, initiated its mobilization plan. Because of the Franco-Russian treaty, German troops were to go to both the Russian and French borders. They would also go to the Belgian border. Count Alfred Schlieffen, onetime head of the Great General Staff, had drawn up a plan for knocking out France by invading Belgium.

At the last minute, Wilhelm decided he did not want a two-front war. His ambassador to London had advised him that Britain would remain neutral if France were not attacked and would guarantee France's neutrality. The kaiser called for the current chief of the Great General Staff, Helmuth von Moltke the younger, grand nephew of the "great" Moltke. He ordered the army chief to cancel troop movements to the west because he wanted to concentrate all of his forces against Russia.

Moltke was shocked. "Your Majesty, it cannot be done," he said. "The deployment of millions cannot be improvised. If Your Majesty insists on leading the whole army to the East, it will not be an army ready for battle, but a disorganized mob of armed men with no arrangements for supply. Those arrangements took a whole year of intricate labor to complete."[2]

Actually, there was a "plan B," for a mobilization only on the Russian border, but the two-front mobilization had already begun. Some 11,000 trains were on the move. Each train was following a precisely timed schedule, passing checkpoints it must reach at the exact minute called for in the plan. Any change in the plan now would be like thrusting a screwdriver into the moving works of a clock.

Wilhelm was going to get a two-front war whether he wanted it or not, and the world would get the most murderous war in history up to that time. For this, blame or credit must go to General von Moltke's great uncle.

The Chief of Staff Takes Over

The Prussian military had never been fond of railroads before Helmuth von Moltke the elder. Prussian generals remembered the maxim of Frederick the Great that more and better roads would only facilitate invasion.[3] What was true of roads must also be true of railroads. In the 1850s, Prussian use of railroads was clumsy. Change came with the appointment of Moltke as chief of the Prussian General Staff. At the time, the General Staff was merely an advisory body to the king, who traditionally gave the commanding general a free hand. But Prince Regent Wilhelm, later King Wilhelm I, wanted to command armies, as his ancestor, Frederick the Great, had. The only trouble was that Frederick was a born general and Wilhelm was not. So he relied on his staff, particularly his chief of staff, Helmuth von Moltke. To speed things up, he allowed the chief of staff to give orders to generals in the field.

Moltke had always been interested in railroads. There was no other means of cross-country transportation that could even compare to trains for speed and load carrying capacity. Handling trains, though, required careful planning. One train could not pass another on a railroad, as carriages could pass each other on a road. A slower train could pull into a siding, but sidings were few and far between. As a result, the speed at which trains traveled must be strictly regulated. The problems of scheduling trains in the civilian world led to the establishment of precisely delineated time zones. Switches had to be set at the right times to accommodate trains on different routes.

Moltke wanted a system that would, when mobilization was announced, pick up reserves in various parts of the country and assemble them into regiments where they would receive uniforms and equipment transported from other parts of the country. The newly minted (actually reminted) soldiers would then be tranported to other parts of the country and incorporated into divisions and corps. Finally, the units would be sent to preassigned spots on the border, where, at a designated time, they would advance into enemy territory and take assigned objectives. Behind the fighting troops, trains would bring up ammunition, rations, and equipment that had been earmarked for each unit.

Railroads were so important, in Moltke's view, that in 1864 he set up a section of the General Staff to deal exclusively with railroads. Officers in this branch had to have a strong mathematical background. As Martin van Crevald puts it, the railroad section of the General Staff attracted the crème de la crème of the Prussian officer corps. In all previous wars, logistics personnel merely supported the frontline troops. But under the Prussian system, they took over management of war.[4] To prevent a surprise attack, officers of the railroad staff maintained constant surveillance of the rail systems of all neighboring countries. To ensure that Prussia's civilian rail network could easily be adapted to military use, Moltke belonged to the state committee for railroads.

Because mobilization by rail was so intricate, every step had to be precisely calculated and monitored by telegraph and, later, telephone. Troops were mustered and trains given trial runs for years before any incident requiring mobilization occurred. Every move had to mesh perfectly. Some railroad facilities were created that would seem grotesque if only civilian needs were considered. For example, some loading platforms near the Prussian border were a full mile long so that several troop trains could unload at once.[5]

The First Blitzkriegs

The first test of Moltke's system came in the Danish War of 1864. This was the first of a series of wars that Otto von Bismark, the Prussian chancellor, contrived in an effort to create a united Germany with the king of Prussia at its head. The Danish War was fought over the provinces of Schleswig and Holstein, which had a mixed population of Danes and Germans. Bismark induced Austria to join him as champions of a still-disunited German nation and invade Denmark. For once, the Prussian railroad system performed perfectly. The war was all over in six months, with the two provinces becoming a joint protectorate of Austria and Prussia. Moltke's handling of the railroads in this war was what induced Wilhelm I to make him, in effect, commander in chief of the Prussian army.

The settlement of the Danish War gave Bismark an opportunity to quarrel with Austria. The aptly named Seven Weeks' War in 1866 saw Prussia, the smaller of the two antagonists, overwhelm Austria and its allies with superior numbers—another fruit of Prussia's railroad-based mobilization, although, as William H. McNeill points out, the Austrians had a habit of conceding after they lost one or two battles.[6] The victory eliminated Prussia's only rival for leadership of Germany, but it was not a total vindication of Moltke's management of logistics. The Prussians had been able to deliver enormous quantities of supplies by rail. But to take them from the railheads to the troops, they had to fall back on horses. Horses and wagons were much less efficient than railroad cars. So toward the end of the Seven Weeks' War, the Prussians were short of food.

There were fewer problems in the third war, the Franco-Prussian War of 1870, which united the German states (except Austria). In this war, Bismark was inadvertently helped by the French in three ways. First, there was their hysterical reaction to the prospect that a

Hohenzollern—a member of the ruling family of Prussia—might take the throne of Spain. Second, by the hubris induced by the reputation their army had of being the best in Europe. Third, by their gross mismanagement of everything from their railroads to their "secret weapon," the mitrailleuse.

For centuries, a Hapsburg had ruled in Austria and another in Spain. And for centuries, French kings had fought the Hapsburgs, who they believed were encircling them. Bismark counted on French fear of having a single dynasty ruling countries at either end of France. When a coup in Spain ousted the queen, Bismark spoke with the Spanish regent, Marshal Juan Prim. A short time later, Prim offered the crown of Spain to Prince Leopold von Hohenzollern-Sigmaringen. Bismark leaked the offer to the newspapers. Not leaked was Leopold's reply: that he would take the Spanish throne only if the monarchs of Prussia and France had no objections.

Just the fact that the offer had been made drove the French newspapers and public wild. Napoléon III ordered his ambassador to Prussia to demand that King Wilhelm *forbid* Prince Leopold to become king of Spain. The king had already said he did not approve of Leopold becoming king of Spain, and he tired of French badgering. He refused to see the French ambassador any more. He sent a memo to Bismark about the situation. Bismark asked if he could publish it, and Wilhelm gave his permission. Bismark edited the memo to make it sound insulting to the French. Napoléon declared war on Prussia.

Prussia was joined by the other German states, and Moltke's mobilization plan demonstrated its worth. Every German soldier arrived on the border with all his equipment on schedule. Munitions, fodder, food, and draft animals arrived with the troops. Railroad flat cars took Prussia's superior steel breech-loading field guns to the artillerymen.

Not all Prussian equipment was superior, however. The French had a far better rifle, and they had the mitrailleuse. They even had a better rail network. But they hadn't coordinated their railroad movements, and they kept their secret weapon—the mitrailleuse—so secret that few of the troops knew how to use it. Their infantry rifle was good, but their infantrymen were outnumbered because the Prussians got there first with the most.

The French mobilization was just short of chaos. Napoléon had planned to invade southern Germany, but the invasion force never got to the German frontier. He planned an amphibious assault at the mouth of the Elbe River. The troop ships got there, but there were no troops aboard. Napoléon declared war on July 15, 1870. He surrendered on September 1, 1870.

The heads of all the German states recognized King Wilhelm of Prussia as emperor, or kaiser, thus, Germany became one nation. Because of their swift defeat of France, the Germans gained the reputation of being great warriors, a reputation that they have somehow managed to retain, at least in some circles—a remarkable feat for a nation that has lost every war it fought for the last hundred years.

More important, Germany convinced every nation in Europe that he who mobilizes fastest wins. Plans were drawn up to take care of all contingencies. That's why when Russia mobilized, it mobilized on the German, as well as the Austro-Hungarian border, and why when Germany mobilized, it mobilized on the French and Belgian borders as well as the Russian.

And that's why all of Europe—and a good part of the rest of the world—plunged into the bloodiest war ever fought up to that time, a war that set the stage for an even worse war.

13

Out of Africa

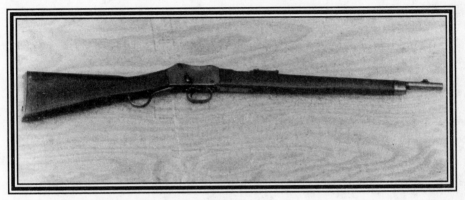

Martini Henry breech loading carbine. Rifles like this were used by both sides in the First Boer War. (Photo by author.)

Bronkhorst Spruit

THE BRITISH COLUMN marching to Pretoria had just reached a stream called Bronkhorst Spruit. The soldiers saw a single horseman approaching from the direction of Pretoria, the capital of what the Afrikaners called the South African Republic and the British referred to simply as the Transvaal, because it was across the Vaal River. The British had annexed the Transvaal because it couldn't pay its debts, but the Transvaalers were now in revolt and were besieging the British garrison in Pretoria.

The approaching rider, a bearded man in an ordinary civilian suit but with a bandolier of rifle cartridges slung over one shoulder, rode up to the British commander, Colonel P. R. Anstruther, and handed him a message. The message was signed by Paul Kruger, Piet Joubert, and Marthinus Pretorius, the triumvirate leading the Afrikaner revolt.

It said any farther British advance would be considered an act of war against the South African Republic. It gave Anstruther two minutes to decide what to do.

"I have orders to march to Pretoria, and that's what I intend to do," Anstruther told the messenger. The messenger told Anstruther that if the column, which consisted of 235 infantrymen of the Connaught Rangers, continued its march, South African troops would resist.

"Do as you like!" Anstruther exploded.[1] More quietly, he asked the messenger to take his reply to the Afrikaner commander and let him know the result.[2] The column waited for the reply without taking any security measures. The British began to see men flitting from bush to bush and from rock to rock. There was movement in the grass. The infantrymen unslung their rifles, and the bandsmen ran back to the wagons to get their weapons. Then a volley of rifle fire blasted out of the veldt, and men dropped all along the British column. The red-coated Connaught Rangers tried to return fire, but there was little they could see to shoot at. The Afrikaners' fire continued like a drum roll, and in a few minutes 120 British, including Anstruther, were dead or wounded. Anstruther, with several wounds and on the point of death, ordered his men to cease fire and surrender. The Afrikaners lost two men killed and five wounded.[3]

The British troops, all veterans of much fighting in Afghanistan, India, and Africa, had met a new kind of enemy. The Transvaal, when independent, could have been described as anarchy tempered by bankruptcy. Its army could be described as a good reflection of the society that created it. There was no standing army, no army reserve, no drill, no training, not even any standard weapons. When danger threatened, the men in a district—there were few towns—organized a "commando" and elected officers. In a big crisis, several commandos might combine and elect a general. Each member of a commando, called a burgher, brought his own horses, rifle, and ammunition. The leader of a commando was called a commandant and commanded anywhere from a couple of hundred to a thousand men. Several field cornets might assist him, each commanding an average of 150 to 200 men. Under the field cornets were corporals who usually led around twenty-five men. The ordinary burgher served under whichever corporal he liked, and the corporals joined the field cornets of their choice. A popular corporal or field cornet might lead twice as many men as his peers.[4] And in action, a burgher would have no compunction against leaving his regular officers and going where he felt he could do the most good.

Burghers received no pay and aspired to no heroics. They considered war an unpleasantly dangerous civic duty. Their main motivation was defense of their homes. An Afrikaner army, then, was a potent defensive force but a rotten instrument for aggression. What made the Afrikaner a potent warrior was the breech-loading rifle and a lifetime of hunting on Africa's plains. Hunting taught him marksmanship and the use of cover and concealment. The breech-loading rifle let him use both of those skills to the maximum. With the breechloader, the Afrikaner could remain hidden as he reloaded, and the rifle's accuracy, combined with his skill in shooting, let him hit game or enemies at ranges the average European soldier would consider impossible.

European soldiers, following practices adopted in the days of smoothbore muzzle-loaders, still fired mostly volleys. They loaded and fired on orders from their officers, although now they made an attempt to aim at their targets. They fired this way partly because they had always done so and partly because their officers were afraid they'd waste ammunition if allowed to fire at will. The Afrikaners did not worry about wasting ammunition. Most of them were poor farmers—the British called all Afrikaners "Boers," the Dutch word for farmer—who hunted for subsistence. They had always tried to make every shot count. If the situation called for a high volume of fire, they could deliver a high volume of fire and do so accurately. Most of the Afrikaners carried single-shot British rifles, the Westley Richards and the Martini-Henry. The Martini was based on an American rifle, the Peabody. In a U.S. Army test in 1866, the Peabody was able to fire seventeen aimed shots and thirty-five unaimed shots a minute.[5] And the Peabody was slower than the Martini, because the Martini had eliminated the Peabody's outside hammer, which had to be manually cocked after a cartridge was loaded. Some of the Afrikaners carried the Swiss Vetterli rifle or the American Winchester—repeaters that could be fired even faster than the two single-shots.

Laing's Nek

British administration in the Transvaal collapsed after Bronkhorst Spruit. All the 1,500 troops there were besieged by Afrikaner commandos. Sir George Pomeroy Colley, high commissioner for South Africa, organized a force to relieve Her Majesty's soldiers. It included 1,200 troops, including 120 sailors, 6 field pieces, and a rocket battery. It started in Natal, on the Indian Ocean, and aimed to cross the Drakensberg mountain range to enter the Transvaal. Colley's column started up a pass called Laing's Nek. Across Laing's Nek and for several miles on either side of it were Afrikaner trenches held by some 2,000 burghers commanded by Piet Joubert.

Colley decided on a combined infantry-cavalry assault on the Afrikaners' left flank. One party of the infantry would stay at the base of the hill crowned by Afrikaner trenches to provide fire support for a second party that would climb a hill so steep it could not be seen by the enemy on the top. Then, while the infantry were holding the attention of Joubert's troops, the cavalry would trot up a gentle slope and overrun the Afrikaner flank.

But the British cavalry started too soon and headed in the wrong direction. They realized their mistake and galloped up the hill to make up for lost time. But they still got there before the infantry assault party was within striking distance. The cavalry's horses were winded by the gallop, and the Afrikaners had no trouble blasting them back down the hill. The Afrikaners, unused to cavalry, were shaken, however. Then the British infantry reached a point where the steep slope turned flat. They formed a line, fired a volley, and charged with fixed bayonets. From their trenches, the Afrikaner riflemen literally blew them away. Of the 480 redcoats who climbed the hill, 150 never came down. The rest stampeded back to the British lines. Afrikaner losses came to fourteen killed and twenty-seven wounded.[6]

The British government now launched a peace proposal: the Transvaalers could have autonomy and their debts forgiven. Colley became worried: if peace came, he couldn't erase the "shame" of his repulse at Laing's Nek.

Schuin's Hoogte

When an Afrikaner force ambushed a British patrol, Colley personally led out a 300-man reconnaissance force equipped with two light field pieces to hunt down the ambushers. It might be the last chance he'd have to beat the "Boers." On a plateau known as Schuin's Hoogte, Colley's men saw horsemen about a thousand yards away. The Afrikaners surrounded the British. There were only 300 horsemen, the same number as Colley's column, but they fired so fast Colley thought there must be 900 of them. The British gunners couldn't get near their artillery without being shot, and the Afrikaner fire was slowly wiping out the British. The British were saved by a torrential rain storm. The Afrikaner commander, Nicholas Smit, was so confident the British could not escape, he let his men scatter to seek shelter from the storm. Colley left his artillery, the dead, and the wounded, escaping with half his force. Afrikaner losses were eight killed and ten wounded.

Peace talks were bearing fruit, and Colley was becoming desperate. He decided on a bold move.

Majuba Hill

At night, he led 600 men to a flat-topped, unoccupied mountain named Majuba Hill. They would organize a base and then bring up what was left of his artillery. Majuba overlooked the main Afrikaner camp. Why they had left the mountain unoccupied, Colley could only ascribe to the "Boers'" total ignorance of things military. He left part of his force at the base of the mountain and brought the remaining 365 men to the top. The move was executed without a hitch. Once on top, Colley assigned sectors to various units, but let the men rest from their climb before digging trenches.

Meanwhile, a woman in a farm house had seen a signal fire Colley's men had lighted to tell those at the British camp they had arrived. She rode over to the Afrikaner camp and yelled, "English! There are English soldiers on Majuba!"[7]

Joubert sent out patrols to prevent British reinforcements getting to Majuba Hill. Then he detailed some troops to shoot at anything they saw moving on the mountaintop. Some eighty young men volunteered to assault the mountain. They began climbing, taking advantage of the naturally terraced slopes to avoid exposing themselves to the defenders. The defenders would not have seen much in any case, because as soon as one peeked over the edge of a cliff, a storm of lead bullets would splatter on the nearby rocks.

The attackers arrived at the last sharp drop-off from the top. One group got below a position called Gordon's Knoll, because it was held by a group of Gordon Highlanders. An Afrikaner corporal gave a hand signal, and the storming party stood up with their rifles in firing position. They fired a volley into the Highlanders at point-blank range, and all but two or three of the Scots were killed. The Afrikaners now held a commanding position from which they could fire at almost all the defenders of Majuba. They concentrated on the nearest defenders, the main body of the Gordon Highlanders. In a few minutes, forty of the Scots were dead. The rest fled in panic. The 80 irregular attackers had climbed a mountain held by 365 regular troops and were driving them back. More Afrikaners from the base camp were now moving up to reinforce them. The Afrikaners began advancing in rushes by two or three riflemen while the rest fired on the defenders to keep their heads down. All British positions on the mountain came under fire. Again, panic seized the British. Colley tried to stop the rout, but an Afrikaner bullet hit him in the forehead. Thomas Fortescue Carter, a war correspondent, later wrote that a "piercing cry of terror . . . rose from the line or group of infantry" and the soldiers threw down their weapons and stampeded down the mountain.[8]

Of the 365 British troops on the mountain, 280 were killed or captured. The British killed one Afrikaner and wounded five others. So much for the importance of drill and the superiority of professional soldiers.

Aftermath

The Transvaal (South African Republic) once again became an independent country. And the British army made a number of changes. One of the most important was replacing the red coats with khaki uniforms. American war correspondent Richard Harding Davis said, "Indeed, the khaki is the English soldier's sole protection. It saves him from himself, for he apparently cannot learn to advance under cover, and a sky-line is the one place where he selects to stand erect and stretch his weary limbs."[9] Another was advancing in open order instead of in dense columns or lines. A third was advancing by rushes. But here, the regular military inclination to do everything "by the numbers" reasserted itself. In training, company commanders would blow whistles to signal the troops when to fall prone and fire, and when to stand up and rush forward.[10]

These changes were to prove valuable—the khaki uniforms, especially—when the British faced the Afrikaners again in 1899. This time, they had to fight both Afrikaner republics: the Transvaal and the Orange Free State. The Afrikaners had some artillery and a few machine guns, and all of them had the superb Model 1893 Mauser rifle. But the British were still using cavalry. About all the cavalry accomplished was the massacre of an Afrikaner unit that was trying to surrender at the very beginning of the war. The failure of cavalry led the British soldier Erskine Childers (later an Irish rebel leader) to write two books, *War and the Arme Blanche* and *The German Influence on British Cavalry*, pointing out the foolishness of relying on swords and lances and the failure of the mounted charge in modern warfare.[11] The books gave Childers an international reputation, but they didn't change regular army thinking until World War I.

14

Battlewagons

U.S. battleships destroy the Spanish fleet as it emerges from the harbor of Santiago de Cuba in 1898. (U.S. Marine Corps copy of 1898 lithograph, National Archives.)

The New Look in Ships

AT THE END of the nineteenth century, battleships and cruisers took on a distinctly modern look. All were powered by steam, and sails were completely gone. All were armored, carried big guns in revolving turrets, and had superstructures like those of today's naval ships. There were still some monitors, mainly for work in shallow water, and there were gunboats and commerce raiders with auxiliary sails, but in general, major naval ships in 1898 looked more like those of 1946 than those of 1865.[1]

The principal fighting ships of all navies were the armored battleships, affectionately called "battlewagons" by U.S. Navy sailors. International trade was increasing in this industrial age and, thanks to the writings of U.S. Navy Captain Alfred Thayer Mahan, world leaders were well aware of the importance of oceans.[2] Britain had the world's largest navy, but Germany, newly united as a result of Helmuth von Moltke's victory in the Franco-Prussian War, thought it needed a navy commensurate with what it considered its importance in the world. A naval arms race began between Germany and Britain, creating still another of the many stresses that led to World War I.

Naval tacticians and strategists spent a lot of time planning how to use these radically different fleets. Tacticians had to revise their thinking because the new ships, unlike the sailing ships of the past, could operate in all kinds of weather. Furthermore, instead of large numbers of short-range guns, they carried fewer, but much heavier guns. And these guns were effective at extreme ranges. In addition to these innovations, there was the "locomotive torpedo." The locomotive torpedo was the gadget we now call simply a torpedo (torpedo originally meaning marine mine). It was invented in 1866 by Robert Whitehead, a Scottish engineer, and Giovanni Luppis, a captain in the Austrian navy. Before the advent of this self-propelled torpedo, torpedoes were either stationary mines, like the ones Admiral David Farragut damned at Mobile Bay or "spar torpedoes." Spar torpedoes were explosives mounted on the end of a long boom that was attached to the bow of a boat. The most famous is the one that the C.S.S. *Hunley* used to sink the U.S.S. *Housatonic* in the Civil War. The blast sank *Hunley*, too, probably because the crew of that Confederate submarine had not closed the hatch, which rode just above the surface of the water. At the end of the nineteenth century, submarines were not yet a major threat to big naval ships, but torpedo boats were. Torpedo boats were small, fast steam launches carrying torpedoes. The torpedoes of that time, powered by compressed air, had a range of only about 500 or

600 yards, but naval tacticians feared that the ponderous battleships wouldn't be able to stop the torpedo boats before they got close enough to launch their missiles.

The naval planners took two approaches. One was to equip the big ships with light, quick-firing guns. One of the first and best was the Hotchkiss revolver cannon—a multibarrel 37 mm gun that looked like a Gatling gun on steroids. The Hotchkiss gun's internal mechanism was quite different from the Gatling's, but it was fired the same way and could throw a stream of one-pound shells at an attacking torpedo boat. Another approach was to hunt down the torpedo boats before they got into range. For this, the naval chiefs created a new class of ship: the torpedo boat destroyer, now called simply the destroyer. Destroyers were bigger than torpedo boats, able to travel anywhere on the high seas. They were faster, too, and they carried heavier guns as well as torpedo tubes. Destroyers were later to become the prime antisubmarine craft. They made torpedo boats much less of a threat, but the little boats didn't become entirely obsolete. John F. Kennedy's PT (for patrol torpedo) Boat 109 was a descendant of those early steam launches.

Naval tacticians dreamed up the ideal battle formation for these new, armored, coal-burning battle fleets. They would engage the enemy in a way that let the ships fire broadsides. This usually means the enemy fleets would steam on parallel courses, anywhere from a few hundred yards to twenty miles apart. But if possible, a fleet would try to "cross the T" of the enemy fleet so that all its ships could use their guns but only forward guns of the first enemy ship could reply. The big battleships would be the backbone of a battle fleet. Between them and the enemy fleet would be the "screen"—cruisers and destroyers, whose guns were lighter and had less range, but that were faster and able to stop any enemy torpedo boats or destroyers.

In the 1890s, these theories of naval combat were still untested. But in the Pacific, at the turn of the twentieth century, two battles took place that would confirm most of the theories and shape naval thinking through World War I. They would mark a turning point not only in naval warfare, but also in world history. They would catapult two new major powers onto the world stage and set the scene for the greatest of all naval conflicts. The first of these battles was triggered by the strategic needs of the new type of battle fleet.

One strategic necessity was refueling. All the major naval powers were interested in obtaining coaling stations on distant shores. This was particularly true of powers with overseas empires like Britain, France, and, to a lesser extent, Germany. Countries without coaling stations had to rely on ships called colliers. Later in the twentieth century, navies switched from coal to oil, but the need to be able to refuel the ships remained. The navies of continental powers, like the United States and Russia, were limited because they had no far-flung refueling stations. When ships relied on the wind, they could operate anywhere in the world almost indefinitely. But now the U.S. Navy, for example, could have trouble operating for an extended period off the coast of China. As the United States was a Pacific power with strong trading interest in China, this was a handicap many U.S. officials felt strongly.

Among them were President William McKinley and Theodore Roosevelt, his brash young assistant secretary of the navy. The increasing friction with Spain over treatment of its colony, Cuba, looked like an opportunity to Roosevelt and the clique he led. One member of his clique was Captain Alfred Thayer Mahan, the world's leading naval theoretician, who stressed the need any major power had for overseas coaling stations. Roosevelt saw opportunity if war broke out between Spain and the United States because one of Spain's colonies was the Philippine Islands. The Philippines could provide a secure base for U.S. ships, which were now operating out of Hong Kong, a British colony. If war did break out, U.S. ships should immediately head for Manila and attack the Spanish fleet. Meanwhile, Roosevelt would do all in his power to make sure that would happen.

Manila Bay

One day when Roosevelt's boss, Navy Secretary John D. Long, was absent, Assistant Secretary Roosevelt appointed Commodore George

Dewey commander of U.S. naval forces in the Far East, the "China Squadron," based in Hong Kong. Dewey, an elderly firebrand, was known for his eagerness to do battle. He was also a member of Roosevelt's clique. In the event of war, he was to proceed to the Philippines. Roosevelt was pretty certain that there would be such an event.

One reason was President McKinley. McKinley was an amiable soul who had the knack of getting people to push him in the direction he wanted to go. But the apparently spineless McKinley knew exactly where he wanted to go, and—with one exception—he always got there. He wanted war with Spain. He got it. He wanted the Philippines. He got them, too, and also Puerto Rico, Guam, and Hawaii—the last the result of a phony revolution that smelled so bad his predecessors had refused to touch the islands for a decade. McKinley was, according to Henry Adams, the "first genius of manipulation."[3] The one thing he didn't get was Cuba. He had let sympathy for the Cubans get out of hand, which was understandable, considering the atrocities committed by General Valeriano Weyler y Nicolau, the Spanish governor. Into the bill authorizing American intervention in Cuba, Senator Henry M. Teller of Colorado added an amendment requiring that the United States would not exercise "sovereignty, jurisdiction or control" over the island. There were no amendments concerning Puerto Rico, Guam, or the Philippines because nobody even thought about those territories.

McKinley had begun sending Spain demands that it modify its conduct in Cuba. The Spanish quibbled and procrastinated, but eventually gave in. And each time they did, McKinley would say Spain's compliance was too little and too late. Then he'd make another demand. Anti-American feeling grew in Cuba among the *Peninsulares*— people who had been born in Spain. Things got so nasty that even McKinley's rabidly anti-Spanish counsel-general in Cuba, General Fitzhugh Lee, advised the president that sending an American naval ship to Havana would aggravate relations with Spain. So McKinley sent the battleship U.S.S. *Maine* to Havana. There was, as everyone knows, an explosion, and *Maine* sank with heavy loss of American lives. Roosevelt was outraged when the navy's top ordnance officer said he knew of no mine that could cause as much damage as *Maine* suffered. (These days, the consensus among experts is that spontaneous combustion in one of the ship's coal bunkers ignited a powder magazine.)[4] But the public was already angry at Spain and believed that the Spanish government, which had the most to lose from the explosion, was responsible. Almost everyone was in favor of taking action against Spain.

When war began, Dewey and the China Squadron sailed from Hong Kong. The China Squadron was not exactly an imposing fleet: it had nine ships, none of them the newest in the U.S. Navy although all were steel, three were cruisers, one was only a revenue cutter, and there were no battleships. The Spanish had forty naval vessels in Manila Bay. The Americans' British hosts in Hong Kong were not optimistic.

"The prevailing impression among even the military class in the colony was that our squadron was going to certain destruction," Dewey later wrote. "In the Hong Kong club it was not possible to get bets, even at heavy odds, that our expedition would be a success, and this in spite of a friendly predilection among the British in our favor. I was told, after our officers had been entertained at dinner by a British regiment, that the general remark among our hosts was to this effect: 'A fine set of fellows, but unhappily, we shall never see them again.'"[5]

Dewey's ships entered one of the two channels into Manila Bay at night, past the fortresses of Corregidor and El Fraile. This channel, the widest, was only four miles wide. The channels were supposed to be mined, but Dewey had served under David "Damn the torpedoes" Farragut. He sailed straight through. His squadron was almost through when some soot in the funnel of *McCulloch*, the revenue cutter, caught fire. Spanish gunners on El Fraile fired at the flash. *McCulloch* and the cruiser *Boston* fired back. The El Fraile guns did not reply. Dewey wasted no time shelling the Spanish fort. He was after the Spanish fleet.

Patricio Montojo y Pasaron, the Spanish admiral, did not rate his fleet nearly as high as the British in Hong Kong. He had forty ships, but most of them were wooden gunboats, utterly worthless in a fight against modern naval ships. His cruiser *Castilla* had developed a leak in its propeller shaft while at sea. The only way the Spanish sailors could save the ship was by plugging the leak with cement. *Castilla* could not move under its own power. The rest of his fleet was more or less decrepit. Dewey's ships were in the harbor before Montojo learned that the four 150 mm guns he thought were guarding the entrance were not in place and that only five of the fourteen mines available in Manila had been placed in the channel. Montojo had no illusions about the outcome of the battle. He had anchored his ships close to shore so his sailors would not have far to swim.

When Dewey was a little more than three miles from the Spanish fleet, he told Captain Charles V. Gridley of his flagship, the U.S.S. *Olympia*, "You may fire when ready, Gridley."

The American ships firing broadsides passed the line of Spanish. The Americans turned in the opposite direction and fired more broadsides. They passed and repassed the Spanish ships five times. Then, because they did not have the newly developed smokeless powder, the Americans pulled back to let the smoke clear and assess the situation. The American ships sustained very little damage, while the Spanish fleet was devastated. Only one ship, *Don Antonio de Ulloa*, was still afloat. The Americans returned to the battle and sank it. The Spanish naval base hoisted the white flag. The Americans lost no ships and no sailors killed and six were slightly wounded in the battle. The Spanish had 300 killed and 400 wounded.

"Mein Gott, Mein Gott!"

Back in Hong Kong, Prince Henry of Prussia, the kaiser's younger brother, had teased Dewey. "Well, Commodore, good luck," he had said. "I may send some ships to see that you behave."

"I should be delighted to have you do so, Your Highness," Dewey replied, "but permit me to caution you to keep your ships from between my guns and the enemy."[6]

Henry, commander of the German Asiatic Squadron, had, a few days after, cabled Berlin that "the natives [of the Philippines] would gladly place themselves under the protection of a European power, especially Germany."[7]

Where Henry got that information from is not known, but it certainly wasn't from any Filipinos, a people who valued their independence enough to fight the American occupiers for years. At any rate, he sent some German warships to Manila Bay. The Germans had more tonnage and more guns than Dewey, and they behaved in what Dewey considered a rude manner. He was blockading Manila, but the Germans refused to identify themselves or to heave to for boarders. Finally, Dewey had had enough. One day *McCulloch*, the revenue cutter, signaled to the German cruiser *Cormoran*, "I want to communicate." *Cormoran* ignored the signal and steamed away. *McCulloch* fired a shot across its bow. *Cormoran* heaved to and was boarded. The next day, Otto von Diedrichs, the German admiral, sent an officer to Dewey to protest.

An officer on *Olympia* told John Barrett of the *New York Journal* what happened:

> The Admiral [Dewey had been promoted right after Manila Bay] has a way of working himself up to a state of great earnestness as he thinks out a question. Commencing in a subdued tone, he gradually became querulous and then emphatic as he spoke of the activity of the Germans. Growing more earnest, his voice took a higher pitch until he complained in vigorous terms of what had been done:
>
> "If the German government has decided to make war on the United States, or has any intention of making war, and has so informed your Admiral, it is his duty to let me know."
>
> Hesitating a moment, he added:
>
> "But whether he intends to fight or not, I am ready."

The German officer left Dewey's presence murmuring, "Mein Gott, Mein Gott!"[8]

There was also a British squadron in the bay, and Britain was the only country in Europe to side with the United States in the Spanish-American War. Neither von Diedrichs nor Prince Henry wanted to start a world war because of a crazy old man in Manila Bay. There were no more discourtesies from the German ships.

Tsushima Strait

A half-dozen years after Manila Bay, Russia was busy trying to establish a "sphere of influence" in the Chinese province of Manchuria. At the same time, Japan was trying to establish one in Korea. They clashed, and on February 8, 1904, Japanese admiral Togo Heihachiro attacked Russian ships in the Manchurian port of Port Arthur. Two days later, Japan declared war.

The declaration caused considerable amusement in Russia. A little nation of "little yellow people" was declaring war on the largest nation in Europe. Russia had an enormous army, although most of it was thousands of miles from Manchuria, with only a single railroad track connecting Europe with the Russian Far East. But Russia also had an enormous navy—more battleships than any country except Britain and France. When what the Russians called "little yellow monkeys" landed in Manchuria and surrounded Port Arthur, the Russians realized how important their navy was. On August 10, Russia tried to move its Far East Fleet out of Port Arthur. A smaller Japanese fleet defeated it and drove it back into the besieged harbor.

Russia's Black Sea Fleet was forbidden by treaty to pass the Dardanelles while Russia was at war, but its Baltic Sea Fleet had no such restriction. It set out for Manchuria in a most inauspicious manner. Russian intelligence reported to Admiral Zinovi Petrovich Rozhentvensky that there were Japanese torpedo boats in the North Sea. In the fog, Russian sailors thought they saw some of those boats and opened fire. A short while later, they realized that they sank a British fishing boat. Only a hasty apology saved Russia from war with the United Kingdom as well as Japan.

Trigger-happy Russian gunners later fired on Swedish, French, and German commercial ships they mistook for Japanese before they reached Tangier to pick up fuel and supplies. Leaving Morocco, they snagged a telegraph cable, cutting off communications between Europe and Africa for four days. Sailing through the unaccustomed heat of the tropics as they headed for the Cape of Good Hope, Russian sailors fell ill by the score. The Russians stopped at Madagascar to wait for a supply ship from Russia. They needed more shells to make up for the ones they had fired at inoffensive neutrals. When the supply ship pulled into the steamy harbor in Madagascar, the Russian sailors found, instead of shells, 12,000 winter coats and 12,000 fur-lined boots.

The fleet was still at Madagascar on March 10, 1905, when the Japanese decisively defeated the Russian troops at Mukden. For all intents and purposes, the war was over. But nobody changed Rozhetvensky's orders. He proceeded to the Sea of Japan.

Togo was waiting. His ships were newer and faster than Rozhetvensky's, but the Russian ships had better armor. The Russian admiral was no fool. He knew the Japanese were probably waiting on the other side of Tsushima Strait, between Japan and Korea. He ordered his captains to form a line abreast instead of a column. That way, if Togo attempted to cross his T, the ships could make a quarter-turn and match broadsides with the Japanese. Even better, if they encountered the Japanese in a column, they could cross Togo's T.

Unfortunately, the Russian navigators could not form a line abreast. The best they could do was form two parallel columns. At 1:40 P.M., the faster Japanese ships approached from the side and performed a double T crossing. Of major ships, Togo had four battleships and eight armored cruisers. Rozhetvensky had seven battleships and nine armored cruisers. But the Japanese were the T crossers. They concentrated their fire on the two lead Russian battleships. One was knocked out immediately, the second twenty minutes later. A third Russian battleship caught fire. Togo's flagship was hit several times, and one of his cruisers was forced to drop out of line while the Russian fleet tried to turn and nullify the T crossing. The Japanese turned with them.

Seriously wounded, Rozhetvensky was transferred to a destroyer. When he returned to consciousness, he was in a Japanese hospital. He learned that five of his seven battleships had been sunk. Three of his cruisers fled to the Philippines, where their crews were interned. One cruiser and two destroyers reached the Russian port of Vladivostok. Altogether, the Russians lost thirty-four of thirty-seven warships. The

Japanese had three cruisers damaged and three torpedo boats sunk. Of the sailors, 4,830 Russians were killed, 5,917 captured, and 1,862 interned. Just 110 Japanese sailors were killed.

Russia had no choice but to sue for peace. Soon after the war, a revolt broke out in Russia. It was put down, but it was a precursor of a successful revolt that would occur twelve years later. U.S. president Theodore Roosevelt mediated the treaty, the terms of which were, of course, favorable to Japan. Roosevelt would have preferred that the Russian defeat had not been so crushing. Ideally, the war would have checked two potential U.S. rivals in the Pacific. This was not to be. And for the next four decades, American and Japanese naval strategists planned for what they considered an inevitable conflict with each other in the Pacific.

15

The Meat Grinder

U.S. soldiers man a portable 37 mm gun to destroy German machine gun nests during an advance in 1918. (U.S. Army photo, National Archives.)

Men Against Machine Guns

THE MEN WERE BRAND NEW—members of what the British called the "New Army" or "Kitchener's Army"—men recruited to make up for the heavy losses suffered by the British regulars during the German offensive of 1914, the "race to the sea," and the strange trench warfare that followed. They were about to get their first taste of battle in the Somme area.

It was July 1, 1916. The British commander, Sir Douglas Haig, had a plan to break through the German line and end the mostly subterranean warfare that had been going on for a year. Haig's offensive had begun with an unprecedented artillery bombardment. He had concentrated 1,000 field pieces, 182 heavy, long-range guns, and 245 heavy howitzers behind the British line. For the last week, the artillery had been dropping 3 million shells on the German trenches and artillery positions. Where the German trenches had been, there was nothing but churned-up earth and shell holes in shell holes. It looked as if nothing could have survived there.

To make sure nothing would, the artillery would lay down a "creeping barrage" in front of the advancing infantry. It would shell the nearest German trenches until the Tommies got close, then it would pound positions farther back. The British officers envisioned the attack as a stroll across the cratered no-man's-land and into what had been the enemy trenches, one line after another until they broke the German line. Then the cavalry, Haig's favorite arm, would charge through and disrupt the enemy's rear areas.

Haig made some changes in traditional tactics, because these "New Army" men were not like the highly trained regulars he had been used to commanding. The "advance by rushes," with troops alternately dashing forward and falling flat to fire and cover the rushes of other troops, would not be done today. Regulars had been trained in that technique, with each rush and pause to fire signaled by whistle-blowing officers. The British had learned in the Boer Wars that advancing by rushes greatly reduced casualties, but made it harder to control the troops. Haig and his staff were afraid that some of these green troops would flop into shell holes and neglect to come out. Instead of advancing by rushes, the Tommies would walk across the field in a somewhat open line. In this case, that didn't seem to be very dangerous. Considering the pounding the Germans had taken, there didn't seem to be much chance of heavy resistance.

In fact, though, the artillery barrage did not do everything it was expected to. It did not cut the barbed wire entanglements in front of the German trenches. It just tossed the wire up and tangled it some more. It was now even harder to get through. That was not a major problem, however.

Few of the Tommies even reached the wire.

The German trenches had dugouts up to thirty feet deep. The most secure were used to house their Maxim machine guns during heavy shelling. When the shelling stopped, many of the Germans were able to drag the machine guns out of their dugouts.

"When the English started advancing, we were very worried," a German infantryman recalled later. "They looked as if they must over-run our trenches. We were very surprised to see them walking. . . . When we started firing we just had to load and reload. They went down in their hundreds. You didn't have to aim; we just fired into them."[1]

The machine guns fired toward their flanks, so a single gun could catch scores of men in its cone of fire. One Maxim gun virtually wiped out two British battalions. Many of the attackers were killed right in front of their own trenches. Long-range machine gun fire killed many troops from reserve trenches before they even reached the *British* frontline trenches. On that first day of the Battle of the Somme, 100,000 troops went "over the top." And 20,000 of them died.[2] Another 40,000 were wounded, and many of the wounded died soon after. Many entire wars have been fought with fewer casualties than what the British army suffered in that one day of fighting.

The casualties did not deter Haig, however. He kept his offensive going for more than four months. The artillery bombardments were able to practically obliterate the German trenches and to kill huge numbers of Germans, but they couldn't kill them all or silence all the machine guns. A few Maxims cut down thousands of infantry-men struggling across the ravaged no-man's-land. For the first two weeks, the British didn't gain an inch, although Haig continued to throw foot-sloggers at the machine guns. Eventually, the British gained a little ground, but gaining ground was useless without a break-through, and there was no breakthrough.

In mid-September, the British unveiled a terrifying new weapon: the tank. Thirty-six of these armored monsters lumbered across the

The Common Touch

Sir Douglas Haig had a well-earned reputation as a cold fish—someone with no empathy whatever for the men serving under him. On occasion, though, he did appear in the trenches and try to make small talk with the troops. He noticed a soldier engaged in one of the routines caused by the unsanitary conditions of trench warfare—removing body lice from his shirt.

"Picking them out, I see," said Sir Douglas.

"No sir," replied the Tommy. "Just taking them as they come."

battlefield and gained 3,500 yards—the longest one-day British advance of the battle—before mechanical breakdowns halted their attack. With many more tanks, more extra parts, and facilities for retrieving tanks, the result might have been different. But in spite of the tanks, the battle dragged on until November 19. By that time, the British army had advanced seven miles and gained a lot of utterly useless real estate. To get that, it lost 419,654 men.[3] That's more dead than the United States lost in *all of World War II.*

What's even more horrifying is that the Battle of the Somme was not unique. Verdun, Ypres I, II, and III, Tannenberg, and dozens of other battles, large and small, were proportionately as bloody. The British Empire lost a million soldiers in what some still call the Great War. France lost 1.7 million men defending their country, which should give some pause to the half-wits who sneer at "cheese-eating surrender monkeys." Germany lost 2 million, which averages out to a million for each of the two fronts the kaiser sought to avoid at the last minute. Austria-Hungary lost 1.5 million. Russian dead came to 1.7 million, and the Italians, fighting on a much shorter front, lost 460,000. Nobody knows how many Turks died in the war, but it had to be close to a million.[4]

Many more people died in World War II, but most of them were civilians. There never was and never has been such a mass slaughter of soldiers as World War I.[5] It introduced many weapons that were to change warfare, such as the tank and the airplane; fearsome weapons like poison gas, which though seldom used later, changed the way war would be conducted in the future; and weapons like the flamethrower, which though also pretty fearsome, are still being used. The Great War left its mark on world history as well as military history. The strain broke the Russian army and caused the Communist Revolution. It caused massive mutinies in the French army that came close to collapsing the Republic. It caused mutinies in the British army, which just avoided a huge mutiny like the one the French suffered. It temporarily broke the Italian army and caused the rout of Caporetto. Finally, the German army fell apart and Germany was wracked by civil conflict. The German collapse led to the myth that the army had been "stabbed in the back," which led directly to Adolf Hitler and World War II.

The Roots of the Slaughter: Mobilization and Weapons

There are several reasons for this horrendous slaughter. One, of course, is the type of mobilization Moltke the elder introduced in the Seven Weeks' and Franco-Prussian Wars. Every nation was convinced that if it could get all its men to the front in time, any war would last no longer than those two nineteenth-century Prussian conflicts. This line of thinking was responsible for the wild optimism seen among both military personnel and civilians. General Sir Henry Wilson wrote about the British plan to seek volunteers for the New Army that Kitchener's "ridiculous and preposterous army of twenty-five corps is the laughingstock of every soldier in Europe. . . . Under no circumstances could these mobs take the field for two years. Then what is the use of them?"[6] Wilson was sure the war would be over long before that. In Paris, Berlin, and every capital of a belligerent state, women threw flowers and men cheered as the troops marched off to their troop trains. It almost seemed as if the soldiers were going on a picnic.

Another glaring reason for this slaughter is the new, or somewhat new, weapons that appeared. The automatic machine gun was just one of these weapons, although by far the most deadly. It has been estimated that just one type of machine gun, the Maxim, has killed more human beings than any other gun in history. Automatic machine guns had been used in a number of previous wars—the Spanish-American War (to a small extent), the Philippine Wars, the Second Boer War, and the Russo-Japanese War—so it wasn't exactly a surprise weapon. The magazine rifle is even older than the automatic machine gun, but during World War I it, too, was a notable manslayer.

One type of surprise weapon was the giant siege guns Germany and Austria-Hungary used to crush the most modern fortresses. The forts, largely subterranean, were designed to withstand the heaviest artillery horses could draw. The German 420 mm Krupp howitzer, "Big Bertha," and the Austro-Hungarian 305 mm Skoda mortar were not drawn by horses. They were drawn by tractors, usually after having been broken down into three pieces. The giant guns destroyed the best steel-concrete-and-earth forts on both the western and eastern fronts. But at Verdun, they were stymied by the trenches and dugouts the French constructed beside their ruined forts. The big guns did start a new trend in warfare that extended into World War II. During the siege of Sevastopol in 1942, the Germans bombarded the city with an 800 mm (31 inch) monster called Dora.

Field Marshal Erich von Manstein wrote that Dora was "a miracle of achievement. The barrel must have been 90 feet long and the carriages as high as a two story house. Sixty trains had been required to bring it into position along a railway specially laid for the purpose. Two antiaircraft regiments had to be constantly in attendance."[7]

Just getting Dora into position was a major enterprise, but the Germans built a couple of 600 mm howitzers in World War II, Karl and Karl II, that were actually self-propelled. The Germans had plans to build another monster with a range of 118 miles to bombard England—a modern version of their famous "Paris gun" of 1918—but that never got off the drawing boards.

Although in World War II the United States built a muzzle-loading thirty-six-inch mortar called Little David to crack Germany's Westwall, super huge guns never became a major trend. It was easier and cheaper to do anything they could do with bombing planes. And the American forces had penetrated the Westwall before Little David got to Europe.

Far more important than the big guns of World War I was a little gun: the trench mortar. Trench mortars are a mainstay of every army in the world and most guerrilla outfits, too. Like the hand grenade, an ancient weapon that was revived in World War I, trench mortars have greatly outlived the trench warfare that created them.

The true quick-firing field gun also got its first workout in World War I. The quick-firer used a brass shell case, like the case of a rifle cartridge, for faster loading, but its great improvement over all previous artillery pieces was its recoil-absorbing mechanism, first applied with total success to the French 75 mm M 1897 field gun. In its most basic form, this consisted of two connected cylinders, one filled with oil and the other with gas or air. When the gun fired, it recoiled in its cradle. This motion forced back a piston in the oil-filled cylinder, which pushed the oil through a narrow opening into the gas-filled cylinder, thus slowing the recoil motion. The oil entering the second cylinder pushed back a rod-less piston, which compressed the gas or air. At the end of the recoil stroke, the compressed gas or air reasserted itself and forced the oil back into the first cylinder, which brought the gun proper back into firing position. As a result, with a well-balanced recoil system, a field gun never moved from its firing position. In all the centuries before 1897, whenever a mobile field gun fired, it rolled back and had to be manhandled back to its firing position. The lack of movement of the quick-firing gun was greatly appreciated by the gunners, and not only because they didn't have to push the gun back after each shot. It was now possible to put a bullet-proof shield on the gun so the gunners could shelter behind it. With older artillery pieces, standing behind the gun when it fired could be suicidal. When it was able to catch enemy troops in the open on relatively level ground, the French 75 was a great killer. On hilly ground, the German howitzers, now also equipped with efficient recoil mechanisms, proved superior to the French field guns, because they were capable of indirect fire—

that is, hitting enemies hidden behind hills. World War I set a pattern for artillery that is still followed.

It also produced a huge number of weapons designed for trench warfare that have continued in service long after the war that caused their development. A prime example is the submachine gun (SMG), a weapon that looks like a short rifle and fires pistol ammunition, but fires it automatically. The first of the type was the Italian Villar Perosa, which was more like two tiny machine guns mounted together that fired pistol ammunition than a typical SMG. The German Bergmann, the second SMG, is more typical. So is the third, the Thompson SMG, which its developer, General John Taliaferro Thompson, called a trench broom. The Thompson and most other SMGs didn't get produced in time to make World War I, but they abounded in World War II—and became obsolete soon afterward.

Poison gas, both delivered in artillery shells or released from cylinders to float across the battlefield on the wind, was a new weapon, although the ancients had occasionally used stink bombs. Gas was a weapon that inspired terror—justifiably. A Signal Corps veteran of World War I once told me that gas was his greatest fear. He operated a telephone switchboard in a dugout below the trenches. It was impossible to be a telephone operator while wearing a gas mask. The veteran explained that if a gas shell exploded near the trench above his dugout, the gas, being heavier than air, would flow down into the dugout before the operators would be likely to hear the gas alarm. And a gas mask was a long way from full protection. Mustard gas and Lewisite could cause terrible burns on any unprotected skin. If inhaled, they burned the lungs. Those not killed by gas were often crippled for life by it. After the war, all industrial countries began building up huge stocks of gas—not only the World War I standbys (chlorine, phosgene, and mustard) but the new nerve gases that could kill just by touching exposed skin. But in spite of these stockpiles, poison gas was seldom used after World War I. When it was used, as it was by the Japanese against the Chinese or the Iraqi army against the Iraqi Kurds, it was against foes who could not reply in kind. The situation was a kind of precursor to the "balance of terror" that has governed the nonuse of nuclear weapons since World War II.

The Roots of Slaughter:
Tradition and Willful Ignorance

The American Civil War had demonstrated the enormous lethality of muzzle-loading rifles. The two Boer Wars had shown how that lethality was magnified by breech-loading and repeating rifles. The Russo-Japanese War gave gory testimony to the power of the automatic machine gun, as well as of high explosive shells. But the European general staffs ignored these powerful lessons.

Haig and his staff officers in the Battle of the Somme were not alone in opposing machine guns with idiotic tactics. British troops at least had khaki and olive drab uniforms, which tended to blend into the background. German troops wore "field gray," a greenish gray that had the same effect. But French infantry wore bright red trousers and cavalrymen retained their polished brass helmets at the beginning of the war. After a few thousand Frenchmen had been killed, the red pants were replaced by "horizon blue" pants, matching the tunics. Some conservatives were still outraged. "Red pants are France!" they cried.[8]

German infantry tactics made even Haig's open-order "stroll" look brilliant. The German infantry went into action in close order as prescribed by the Drill Book of 1888. They had used similar tactics in the Franco-Prussian War of 1870 and watched men cut down by the score by enemy troops with breech-loading rifles. Nevertheless, in the 1880s German military authors warned that troops would take advantage of their distance from their officers to hide, or even desert, causing the failure of attacks in open order. They agreed that close order would result in more casualties, but they thought this was a lesser danger than loss of control by officers.

This desire for total control led foreign observers to two conclusions: either the German high command was incompetent or German soldiers were untrustworthy. Lieutenant Carl Reichmann, a U.S. Army observer of German maneuvers in 1893, reported, "They evidently intend to handle their infantry in close lines in the next war. The average German private is not a person to be turned loose in a skirmish line and left to a certain degree to his own devices. . . . They prefer to lose men than to lose control of the officers over them."[9]

The most popular formation for attacking infantry was a "column of platoons," a formation Prussian soldiers had used since the Napoleonic Wars. The column would have a front of about twenty-five meters (about twenty-seven yards). Within those twenty-five meters were six lines of riflemen with thirty-two men in each line. Each line consisted of four German squads; two lines made a platoon. The platoon leader, a lieutenant, marched on the right side of the first line so that all the men would be within the sound of his voice. Behind each platoon marched four of the squad leaders, sergeants, to make sure nobody lagged. In front of the whole column, right in the middle, was the company commander. The troops fired volleys on command. Volleys were supposed to establish fire superiority and make the enemy keep his head down. There would inevitably be losses, but the troops, under the watchful eyes of their officers, would continue on, cowing the enemy with their *Furor Teutonicus*.[10] Faith in the Furor Teutonicus was matched by the French belief in the superiority of the bayonet, wielded by soldiers full of French élan.

Neither furor nor élan cowed any enemies. The German theory was first tested at Mons on August 23, 1914, where some of the best troops of the German army, such as the Brandenburg Grenadiers, attempted to break through a line of outnumbered British regulars holding stone walls and abandoned houses. "Here we were advancing as if on parade ground," recalled a German officer.[11] Then the British began firing. They fired so fast the Germans were convinced that each house and wall concealed a machine gun. Although the Germans had superior numbers and were supported by much superior artillery, the attack failed, and they lost 50,000 men.

Losses like that did not convince the German high command to greatly modify its infantry tactics. In October, during the so-called race to the sea, each side was trying to outflank the other. Seven British divisions were pushing into territory they thought they had reached before the Germans, when they ran into fourteen German divisions. The Germans were volunteers who had not yet received any military training. Among them were thousands of university students who had volunteered as soon as they heard of mobilization. After two months of training, they had been rushed to the front.

"Over every bush, hedge or fragment of wall," wrote a German military historian, "floated a thin film of smoke, betraying a machine gun rattling out bullets."[12] The historian was mistaken. The British and the Germans each exaggerated how many machine guns the other side had. Actually, they had the same number in each of their infantry divisions: two guns to a battalion. The smoke the Germans saw came from rifles.

The results of their fighting in Ypres is known in Germany as the *"Kindermord bei Ypren"* (massacre of the innocents at Ypres). But it wasn't only innocents who were massacred. On November 11, Kaiser Wilhelm came to Belgium, and disheartened by the lack of progress at Ypres, he ordered the eight regiments of the Imperial Guard—regulars and the elite of the German army—to attack at Ypres. The regulars succeeded no better than the hastily trained volunteers. An attack by the First Prussian Foot Guards, the premier regiment of the German army, was blasted back by a collection of cooks and batmen (orderlies for British officers) of the Fifth Company of the British engineers. In the long, bloody battle, the British lost more than 58,000 men. The Germans lost, at the minimum, 130,000.[13] Most of the British casualties were caused by German artillery, which greatly outranged the British guns. The shallow trenches the British occupied protected them from small arms fire, but not from howitzers. The British field guns were outnumbered two to one; the heavy artillery, ten to one.[14] In spite of all that, the British managed to hang on to Ypres.

British military writers credit the superiority of their Short Magazine Lee Enfield rifle for the disparity of casualties between their forces and the Germans in these early battles. The credit is misplaced. The British Lee Enfield is actually less accurate and less powerful than the German Mauser. The British writers contend that because the rifle's locking lugs are not on the front of the bolt, the bolt can be manipulated faster. They also point to the Lee Enfield's ten-round magazine—as opposed to the Mauser's five-round magazine—and say that it promotes speed. Actually, any difference in speed of bolt manipulation between the Lee Enfield and the Mauser would have to be measured in milliseconds, and as both the Mauser and the Lee Enfield are loaded with five-round clips, the Lee Enfield is faster only for the first ten shots.[15]

Clips and Magazines

Clips, sometimes called stripper clips, hold cartridges together so a number can be loaded into a magazine with one motion. Detachable magazines, like those used for the .30 caliber carbine or the .45 caliber pistol, are sometimes called clips, but that is a misnomer. The British Short Magazine Lee Enfield (SMLE) has a ten-round magazine that is loaded with two five-round clips. The Mauser 98 has a five-round magazine that is loaded with one five-round clip. Starting with loaded magazines, the SMLE can obviously fire the first ten rounds faster because no time is spent on reloading. After that, the speed of fire is the same for both rifles.

The reason for the British superiority is simple. They were better shots. As a result of hard lessons learned in the Boer Wars, the British had been training their infantry—even cooks and batmen—in rifle marksmanship. The slowest regular could fire fifteen aimed shots a minute, while the fastest managed to fire thirty aimed shots a minute. British skill with rifles was one reason for the heavy German casualties. The stupidity of German tactics was the other.

Roots of the Slaughter: Isolation of the Generals

In the Middle Ages, King Robert Bruce of Scotland killed an English knight who attacked him while he was scouting the area where the Battle of Bannockburn would be fought the next day. During the American revolution, George Washington himself was in the rifle sights of Patrick Ferguson, the inventor of a breech-loading rifle and the top marksman of the British army. Ferguson did not fire because he thought it was not "sporting" to kill the distinguished-looking officer from ambush. (He did not know it was Washington.) In the American Civil War, Major General Philip H. Sheridan led his division up Missionary Ridge outside Chattanooga and later turned his retreating army around in the Shenandoah Valley and led it against the Confederate army of Jubal Early and crushed it. To put it simply, before armies grew so huge as to require a horde of staff officers, generals *led* their troops.

Giantism in armies did not begin in World War I, and neither did the bureaucratization of leadership. But it reached about the same level it is at now. Lieutenants come under fire today—possibly more than privates. Captains do too. Majors are less often targets of enemy riflemen or machine gunners. Colonels much less and generals very seldom. (We are speaking of armies, here. With air forces, the figures are a bit different.)

What this means is that generals have become isolated from their troops. Henry V could say, "Once more into the breech, dear friends," and lead them into it. Not today. When the commanding general is a Douglas Haig, you have a prime ingredient for the kind of slaughter featured in the Great War. According to historian S. L. A. Marshall, "Haig was as cold as ice. Between him and his troops, there was no bond of sympathetic understanding. But it would be unfair to say that anyone excelled him in the art of getting ahead" in the military bureaucracy.[16] But on both sides in World War I, groups of middle-aged men in headquarters far from the shooting sent tens of thousands of young men out to die then slept soundly in comfortable beds.

Getting out of the Trenches

Trench warfare was depressing for many reasons, but the biggest one was that there didn't seem to be any way to end it. In between the giant offensives like the Somme, both sides engaged in midnight raids on enemy trenches, sniping, and gas attacks. The raids resulted in extremely brutal, close-quarters fighting, with soldiers using such unorthodox weapons as shotguns, daggers, and even clubs.[17] (Shotguns and daggers are still used.) To break the stalemate, the Allies and the Germans came up with two different approaches. The Allies relied on the tank; the Germans, on newly developed "infiltration" tactics that differed radically from the shoulder-to-shoulder attacks they used at the beginning of the war.

The tanks had been introduced before there were enough of them and before they had been made mechanically reliable. Colonel Ernest Swinton, commander of the tank corps and the man most responsible for the tank, opposed their introduction at the Somme, but Haig demanded the new machines. He had lost hundreds of thousands of men, but he thought that tanks would give him his long-sought breakthrough. The first tank had been designed in 1915 when trenches were shallow and narrow and bombardment had not yet turned the battlefield into a churned up morass. Swinton had written a memo outlining a number of conditions—such things as the weather, the state of the terrain, availability of reserve tanks, and so on—that should be met in order to get the most from tanks. The memo was ignored. The Somme almost finished the tank. Haig's staff was full of conservatives who hated the new machine before they even saw it. They gave such a poor report on the performance of the tanks that the War Office cancelled an order for new tanks.

That would have ended tanks in the British army if it had not been for Major Albert Stern. Stern, a reserve officer who was a financial magnate in civilian life, was now in charge of tank construction. When Stern got the order, he went directly to his friend, David Lloyd George, the future prime minister who was now the war minister. Like Stern, Lloyd George thought the order was ridiculous. Stern then went to the chief of the Imperial Staff, Sir William Robertson, and told him the cancellation order would not be carried out. Almost simultaneously with the British, the French had been independently working on another tank design. It took longer to develop, but the French had more faith in their tank. When the British ordered production of their tank, they first called for 40 machines, then increased the order to 150. The first French order was for 400 tanks, then increased to 800.

The tank was not out of danger from its home front enemies. General Staff officers cut back future orders, forbade the installation of radios in tanks—an invaluable means of communications—and refused to let the tanks' cannons fire case shot—the ideal anti-infantry weapon—until the commanders in the field demanded it.

Haig dealt the tank another blow when he demanded tanks for his offensive at the Third Battle of Ypres in 1917. This section of Flanders is almost a marsh at the best of times, but in 1917, all the trees, brush, farmland, and drainage ditches had been shelled into oblivion. When the attack began, the battlefield had been drenched by a long period of heavy rain. The rain continued as the tanks went into action. The tanks were defeated by General Mud. At Cambrai later that year, the tanks were finally used under favorable conditions. Everything went well except in the center of the front. There, the tanks had outrun the Highlander infantry that was supposed to accompany them. And there, the German commanding officer was a General von Walter, an old artilleryman who had taken an interest in tanks. Walter moved some of his field guns close to the front and trained his gunners to hit moving targets. They knocked out eleven British tanks. One German gunner, Sergeant Kurt Kruger, knocked out five of them himself. He was killed when the British infantry finally reached the German lines.[18]

Cambrai was the first tank victory. It might have been a greater victory if so many tanks hadn't been lost in the mud east of Ypres.

Surprisingly, for a nation so technologically oriented, tanks aroused little interest in Germany. Instead, the Germans were relying on new infantry tactics.

All of the armies on the western front had been changing their infantry tactics and developing new weapons to fit the tactics. Infantry advanced now in more open order, with the men frequently taking cover as they advanced. The trench mortar helped them cope with machine guns. So did a small 37 mm cannon, which, like the trench mortar, was light enough to be carried by one or two soldiers and fired from a tripod. The infantry still had their heavy machine guns for defensive use, far more in fact than they had started the war with. But on the offense, they used light machine guns like the Lewis and the Benet-Mercié (the Hotchkiss light machine gun in British terminology). They could answer the enemy machine guns with automatic guns of their own while on the move. The Germans, too, adopted a light machine gun, the Maxim 08/15, though it really wasn't very light. It was just the standard Maxim, water jacket and all, minus the usual "sled" mount and the addition of a buttstock and a bipod. It was heavier, but more reliable than the Allied guns.

No infantry had changed as much as the German. No infantry needed to change as much, either, of course, but the German change was radical, indeed. Most of the details had been worked out on the eastern front, where, for some reason, the brightest German officers, like Max Hoffmann and Oskar von Hutier, seemed to be located. Hoffmann, a gluttonous, unathletic officer, was Erich Ludendorff's brains while Ludendorff was Paul von Hindenburg's brains. Hindenburg was a retired gentleman who was called back to the army and became a popular hero in Germany because he was willing to take advice. Hoffmann's advice to Ludendorff resulted in Ludendorff's advice to Hindenburg, which resulted in the brilliant German victory at Tannenberg. Von Hutier played the major role in developing the Germans' infiltration, or storm trooper, tactics.

These new tactics involved a special assault team that used shorter, lighter rifles, many hand grenades and trench mortars, large numbers of Maxim light machine guns, and flamethrowers. Some of them had SMGs. The assault teams led the regular infantry in attacks. They bypassed enemy strongpoints, leaving them to be reduced by the ordinary infantry, and drove into the enemy's rear areas—an action that seriously affected enemy troops at the front. No longer were there masses of men in the attack. Platoons and even squads were widely separated. Men took cover in shell craters or any other shelter as they moved. Squad leaders and individual riflemen were trained to think for themselves.

The new technique was so effective Russia surrendered. The new technique impressed Berlin so much that Hindenburg and Ludendorff were transferred to the west. Unfortunately for Germany, Hoffmann stayed in the east.

The end of fighting in the east meant that Germany could move masses of soldiers to the west. There was an urgent need for more men in the west because the United States had entered the war. In spite of its enormous industrial capacity, the United States was short of weapons; there had been little attempt to prepare for war. But there was no shortage of men in that vast country. The French and British were prepared to give the Americans artillery, machine guns, and airplanes.

By the time Ludendorff and Hindenburg moved west, all German infantry had been training in storm trooper tactics. Ludendorff decided to use those tactics on a grand scale and break through the Allied line. Swarms of small units, following lines of least resistance, would push deep into Allied territory, wiping out the rear area installations that supported the troops at the front. The Allies would collapse before the Americans could help them.

The Ludendorff offensive opened on March 21, 1918, with a sudden, short, and violent bombardment in the area around Arras. The storm troopers moved through a pea-soup fog and hit the Allied lines. Under Ludendorff's plan, the German Seventeenth Army would deliver the main thrust, break through the junction of the British and French armies, and roll the British line up to the Channel. The Eighteenth Army would act as a flank guard to prevent reinforcements from striking the Seventeenth Army. The Seventeenth Army met unexpected resistance, but the Eighteenth Army drove forward and reached the rear of the British army. If Ludendorff had been really committed to infiltration tactics, he would have reinforced the Eighteenth Army and urged it on. That could have practically ended the war. Instead, he ordered the Eighteenth Army to halt because it had gotten too far ahead. His plan called for the Seventeenth, not the Eighteenth, to deliver the master stroke. After two days, he allowed the Eighteenth Army to resume its advance. But by that time, the Allies had sent reserves to the area and blocked any new advance.

The Germans launched a diversionary attack around Armentières that achieved surprising success. Ludendorff should have quickly shifted massive reinforcements to the area to exploit the gains. Being flexible and taking advantage of opportunities was the basis of infiltration tactics. Instead, the German commander dribbled divisions into the area. By the time it had become a major offensive, the British had brought up more troops and stopped the Germans. Ludendorff continued to pour troops into the stalled offensive but got little more than heavy casualties.

Well into July, Ludendorff kept repeating the pattern. The Germans would make a few gains, the Allies would bring up reserves, and Ludendorff would waste his numerical superiority trying to break through strongly held lines. He even used the few tanks Germany had finally built, but they were outclassed, outnumbered, and defeated by Allied tanks. American troops were rapidly disembarking in France, and the Germans were now outnumbered.

On August 8, the Allies attacked the Germans with 600 British and French tanks in the Somme area.[19]

"August 8 was the black day of the German army in the history of the war," Ludendorff said.[20] It was far from the bloodiest battle of the war, but it was a big one. Casualties came to: French, 24,232, British, 22,202, and German, 75,000.[21] The Allies captured 29,873 prisoners and 499 cannons. More important, it convinced both Ludendorff and Wilhelm that the war could not be won.

Ludendorff and Wilhelm later recovered their nerve. But nobody else did. There were riots and strikes in Germany. Revolution was brewing. Germany's allies were surrendering. First Bulgaria, then Turkey, then Austria-Hungary. On November 11, 1918, Germany surrendered and the kaiser went into exile.

Of all the turning points that appeared in the ground war of 1914–1918—radically new types of artillery, widespread use of automatic machine guns, poison gas, and the rest—the biggest was one that both the Allies and the Germans contributed to but that did not really appear until the next war. The blitzkrieg, which revolutionized war for a few years in World War II and led to changes that are still with us, was based on the tank, the Allies' contribution, and the Germans' infiltration tactics.

16

War Beneath the Waves

German submarine surfaces during World War I. German submarine warfare brought the United States into the war. (Bureau of Ships, National Archives.)

Swimming with the Fishes

IT WAS 1917. The United States had declared war on Germany, but German Admiral Eduard von Capelle, secretary of state for the navy, was not worried.

"They will not even come, because our submarines will sink them," he told a committee of the Reichstag. "Thus America from a military point of view means nothing, and again nothing, and for a third time nothing."[1] The German submarines had been causing the Allies considerable trouble when Capelle was imparting this bit of wisdom, but they were never that effective. In the past three years, they had never been able to shut down shipping to Britain and France. In the next three months, it would become apparent that Germany had lost the U-boat war. The German *Unterseebooten* were not able to keep America out of the war, but they had certainly brought America into it.

In a way, that was a bit ironic. Until comparatively recent years, Germany had no more to do with submarines than it had with tanks. The United States, on the other hand, was the first nation to use a submarine in combat—David Bushnell's *Turtle*, which failed to sink a British warship during the American revolution. And during the Civil War, the Confederate submarine *Hunley* was the first to sink an enemy ship. After the American revolution but before the Civil War, Robert Fulton built a submarine in France for Napoléon Bonaparte. But at the last minute, Napoléon decided that sinking ships while hidden beneath the waves was more like piracy than war, so Fulton's *Nautilus* was never used.[2] Napoléon was a great soldier, but not much of a visionary when it came to weapons. He also disbanded the balloon corps that had been instituted during the French Revolution.

After the Civil War, American inventors continued to design and build models of submarines. The most successful was John Holland, an Irish-born American. Holland, born in Liscannor, County Clare, had been active in Irish revolutionary politics. He was a school teacher in Ireland, but at night he tried to design a submarine that, he hoped, would destroy British sea power. In 1873, he left Ireland to join his mother and brothers in Boston. He resumed teaching school in America and working on his submarine. The Feinian Brotherhood, an Irish-American secret society, financed his work, and he launched a small sub, *Feinian Ram*, in 1881. But Holland and the Feinians quarrelled, and they cut off his financial aid. He built another boat, *Holland IV*, which in 1888 won a U.S. Navy competition for submarine design. He was unsuccessful, however, in selling the boat to the navy. He designed two more subs and sold the second, *Holland VI*, to the navy. The navy renamed it the U.S.S. *Holland*. This design was adopted by many nations, and one of the first was Britain.[3]

Holland took advantage of a couple of devices that had not been

Captain Bushnell and Dr. Bush

Captain David Bushnell of the Continental Army was a brilliant scientist but a very unlucky man. On the night of September 6, 1776, he launched the world's first combat submarine, *American Turtle*. It consisted of two solid, curved pieces of wood, carefully fitted together and sealed against leakage. It had a hand-cranked propeller and a rudder. It would hold one man, who squeezed through a hatch at the top. There was a propeller on the top to force it below the surface, and it could take on water as ballast. Outside the craft was a bomb attached to an auger that could be turned from inside the submarine. The bomb had a clockwork timing mechanism that had already been activated. Both the submarine and bomb had been designed by Bushnell.

Bushnell's first stroke of bad luck was when his brother, Ezra, who had been practicing operating the *Turtle*, became sick the night before the planned launch. A volunteer, Ezra Lee, had taken his place, but Lee was not as skilled a submariner as Ezra Bushnell. Nevertheless, Lee made it to the target, the H.M.S. *Eagle*, flagship of the British squadron in New York harbor. But Lee couldn't force the bomb's screw through *Eagle*'s metal sheathing. The bomb's timer was ticking, and there was no way to stop it. Lee jettisoned the bomb and headed back as fast as he could. The bomb exploded, and the British sailors woke up.

After the war, Bushnell petitioned the Continental Congress for some form of recognition or compensation. But although George Washington strongly praised Bushnell, Congress did nothing. The inventor went to France, where he may have collaborated with Robert Fulton, who was building a submarine for Napoléon. The French sub worked, but at the last minute, Napoléon decided that submarines weren't sporting and cancelled the project.

Bushnell, doubly disappointed, returned to the United
States in 1795 and changed his name. Years later, in 1824,
residents of Warrentown, Georgia, learned that the gentle
Dr. Bush, who taught science and religion at the local
high school, was really David Bushnell, the Revolutionary
inventor.

invented at the time of the Civil War: the internal combustion engine
and the storage battery. On the surface, his boat was powered by an
internal combustion engine burning gasoline. This type of engine was
much smaller and lighter than the steam engines that powered most
water craft at that time. The Confederate David-class submarines of
the Civil War used steam engines with the funnel of the engine pro-
truding above the water, greatly limiting their effectiveness. For underwater travel, Holland used an electric motor powered by a storage
battery. The battery was charged by the gasoline engine while running
on the surface.

Simon Lake was another American submarine pioneer. Lake
designed a submarine for the U.S. Navy in 1892, and launched his first
boat, *Argonaut Jr.*, in 1894. The next year, he formed the Lake Submarine Company, which built *Argonaut*, the first submarine to operate in
the open ocean, in 1898. In 1901, he formed the Lake Torpedo Boat
Company, which built "submarine torpedo boats" for the U.S. Navy
and many foreign navies. Lake invented and patented even-keel
hydroplanes, ballast tanks, divers' compartments, periscopes, and twin
hull design, all of them essential to modern submarines.[4]

Submarines at War

By 1900, six navies had submarines, but Germany was not one of
them. Germany didn't start building subs until 1906.[5] In 1908, it
introduced diesel engines for the surface power, giving the boats
longer range than gasoline engines. Most submarines in those days
were short-range craft, designed for harbor protection. The idea of

using them to attack merchant shipping seems not to have crossed anyone's mind. That was what cruisers were for. Under international treaties and conventions, the attacker would stop the merchant ship, allow the passengers and crew to get off and pick them up if no other refuge was likely, and then either sink the ship or take it as a prize. The Germans even used their long-range engines to build a submarine merchantman, *Deutschland*, which showed up at New York during World War I in June 1916.

Deutschland, however, was more of a stunt than a trend. Almost all submarines were designed to sink other ships—enemy naval ships. Soon after the war began, the Germans sent a flotilla of subs out to sink British warships. *U-15* saw some British navy ships and fired two torpedoes at the battleship *Monarch*. It missed. Then the cruiser *Birmingham* rammed *U-15* and sank it. But the next month, September 1914, *U-21* sank *Pathfinder*, and a few days later, on September 22, *U-9* sank three British cruisers, *Aboukir, Hogue*, and *Cressy*, in the same action. The British decided to step up their submarine production.

On April 25, 1915, Germany sent one of its long-range submarines into action. "Long range" is a relative term. A range of 4,000 miles was long for a submarine in 1915, but hardly so for a battleship, cruiser, or destroyer. The big *U-21* left Kiel for the Dardanelles, in an effort to help Germany's ally, Turkey. It sailed through the Straits of Gibraltar, refueled in Austria, and sank two British battleships, *Triumph* and *Majestic*, in the Mediterranean.

For the first year of the war, it was extremely difficult to locate and destroy submarines. The British conducted patrols around their home islands and the North Sea trying to locate surfaced submarines. A sub in World War I could stay submerged only for a very limited time, but a surface vessel looking for a surfaced sub in the foggy North Sea would spend a lot more time looking than seeing. And only if the sub were seen on the surface was there a chance of destroying it. Finding a submerged U-boat became a little easier in 1915, when the British introduced the hydrophone. This was a passive listening device that could pick up the sound of a sub's propellers. But the sub had to be quite close to be heard, and the ship hunting it had to stop its own engines so it could hear the sub's. In 1916, the depth charge was invented. If a submerged U-boat could be located, it could be destroyed.

This period saw a number of developments in submarine warfare—some practical, some not. Among the latter was the "submarine monitor," which had a twelve-inch gun that could only be fired over the bow: in any other direction, the recoil might capsize the boat. Then there was the super-fast submarine. It had a steam engine that gave it a top speed of twenty-five knots, but it took thirty minutes to submerge.

Minelaying submarines appeared in 1915, taking up a task first performed by the submarine U.S.S. *Alligator* in the American Civil War. Another development was the Q-ship—a merchant ship taken over by the navy with a navy crew and outfitted with hidden deck guns, torpedo tubes, depth charges, and a hold full of lumber to make it harder to sink. The purpose of the Q-ship was to lure surfaced submarines into trying to stop it and then sink the sub. Subs in those days were extremely vulnerable to surface attack. A hole anywhere in the boat would almost invariably put it out of action. The Q-ship was to have a somewhat sinister influence on the course of submarine warfare.

At the outbreak of the war, the Germans had relied on their cruisers to raid British and French shipping around the world. But in less than a year, the modern German commerce raiders had all been taken out of action. The only exception was the ancient *Seeadler*, a full-rigged sailing ship commanded by Count Felix von Luckner. But *Seeadler* preyed only on sailing ships: anything else could easily outrun it.

In 1915, the Germans turned to their submarines for commerce raiding. Under international law, the submarine was obliged to stop the merchantman and allow the passengers and crew to escape. But what if the "merchantman" was really a Q-ship? Many submarine skippers didn't want to take the chance. One such skipper was Lieutenant Walter Schwieger, who, cruising off Ireland, saw the smoke of a steamer in the distance. He ordered a dive, then looked through his periscope. It was a huge steamer, obviously a passenger liner, and in these waters, it was probably an enemy ship. He fired two torpedoes, watched them strike, and left the area.

Schwieger had committed the first of a whole series of strategic errors that were to plague Germany's World War I submarine campaign. The steamer was the Cunard-liner *Lusitania*. It went to the bottom in just eighteen minutes after the first hit, and 1,198 civilians, including 128 Americans, went with it. The American public was enraged. Coming right after word of the German atrocities in Belgium—the shooting of civilian hostages and the burning of towns, cities, and the University of Louvain—the American people were almost ready to declare war on Germany. It seemed only Woodrow Wilson stood between the United States and the war. The American reaction was so strong the German navy cancelled its "unrestricted submarine warfare" for a year. No more passenger ships would be attacked and surprise attacks were forbidden. The center of operations shifted to the Mediterranean, where there were fewer American ships and most of the hundred or so ships sunk were sunk by gunfire from surfaced submarines.

Meanwhile, the British were continuing their blockade of Germany, which did not please U.S. business interests. Secretary of State Robert Lansing sent a note to Britain and France, saying that if the Allied blockade continued, there was no reason the Germans could not continue their submarine campaign against Allied shipping as long as they allowed the passengers and crews of civilian ships to leave before the ships were sunk. Furthermore, the note said, the arming of merchant ships was against international law, and armed ships could then be considered naval vessels.[6]

Fatal Blunders

That note prompted another German blunder. Wilhelm's government decided that the United States was not really against unrestricted submarine warfare. In January 1917, German military leaders argued that if the submarines could sink 600,000 tons of British shipping a month, they could bring the island kingdom to its knees in six months.

On January 9, 1917, the kaiser sent a secret order to submarine commanders authorizing a continuation of unrestricted submarine attacks. British Naval Intelligence intercepted and decoded the order. Then, on January 17 it made another interception—a message from Germany's Foreign Secretary Arthur Zimmermann to the German ambassador in Mexico City. It said:

> WE INTEND TO BEGIN UNRESTRICTED SUBMARINE WARFARE. WE SHALL ENDEAVOR TO KEEP THE UNITED STATES NEUTRAL. IN THE EVENT OF THIS NOT SUCCEEDING, WE MAKE MEXICO A PROPOSAL OF ALLIANCE ON THE FOLLOWING BASIS: MAKE WAR TOGETHER, MAKE PEACE TOGETHER, GENEROUS FINANCIAL SUPPORT, AND AN UNDERSTANDING ON OUR PART THAT MEXICO IS TO RECONQUER THE LOST TERRITORY IN TEXAS, NEW MEXICO AND ARIZONA.[7]

It is really hard to imagine a more stupid move. In normal times, Mexico's military potential compared to that of the United States the way a house cat compares to a leopard. At this time, Mexico was total chaos. The Mexican revolution was still going on, and the country was overrun by revolutionaries, warlords, and outlaws.

The British released both messages to the United States. The German ambassador had already informed the United States of the beginning of unrestricted submarine warfare—on January 31, the day before it began. Zimmermann believed the United States would do nothing, "because Wilson is for peace and nothing else."[8]

On February 3, Wilson broke off diplomatic relations with Germany. On February 23, the United States received a copy of the "Zimmermann telegram." A check showed that it was authentic. The Germans had used American diplomatic cables to send the coded message to Count Johann-Heinrich von Bernstorff, the German ambassador to Washington. American cable files showed that the original message, word for word, had been sent to Bernstorff, informing him of the proposal to be made to Mexico. Berlin had to admit the message was authentic. On April 6, the United States declared war on Germany, which led to the defeat of the U-boats.

Convoys

When the United States entered the war, its army was small and poorly equipped. Its navy was not. It was, in fact, the second largest navy in the world; only the British navy was larger. This had two effects on the war. First, the Allied blockade of Germany, including the North Sea "mine barrage," was greatly strengthened. Not only were fewer materials able to arrive in Germany, fewer U-boats were able to reach the ocean. Second, Admiral William Sims went to England to confer with his British counterparts. Admiral Sir John Jellicoe, the first sea lord, showed Sims the figures on U-boat sinkings. The British were losing ships faster than they could be replaced.

"Looks as if the Germans are winning the war," Sims said.

"They will unless we stop those losses," Jellicoe said.[9] Jellicoe introduced Sims to Vice Admiral Sir Lewis Bayly, who was in charge of antisubmarine warfare. Sims suggested that merchant ships going to England travel in groups escorted by destroyers and other fast warships. He convinced Bayly, then the two admirals convinced their own governments. Soon, U.S. destroyers and other ships were operating out of Ireland, France, and Gibraltar, as well as the United States. The loss rate soon dropped by half, and it became obvious even to Admiral Capelle that U.S. troops definitely would come.

The reason why convoys were effective was not simply that destroyers and corvettes could sink submarines. Convoys spread over an enormous area of sea, and in World War I the ability of escort ships to locate a submerged submarine was limited indeed. There were usually only two or three naval vessels escorting around forty merchant ships. It would be quite possible for a submarine to fire its torpedoes at a ship in the convoy and get away.

Convoys worked because submarines were so slow. On the surface, they were the slowest of all naval vessels. Submerged, they couldn't chase the slowest freighter. So to attack Atlantic shipping, the subs positioned themselves in the seaways and waited for a freighter to pass within reach. If a submarine skipper missed one freighter, another might come along soon after. But with convoys, if a submarine missed a convoy, it was in for a long wait before another appeared. If it did see a convoy and fired, that act would give the accompanying destroyers an idea of where the sub was located, and depth charges would soon follow. Hunting commercial ships had become a lot more dangerous. U-boats began to move out of the high seas and into coastal waters where they could find small, unescorted vessels. But that brought them into range of airplanes and blimps, which could often see submerged U-boats if they weren't down too deep. And they ran into mines. Thanks to American help, the mine barrages in the English Channel and the North Sea had been made much denser.

On October 30, 1918, the German High Seas Fleet was ordered to sail into the North Sea and attack Allied naval ships in order to save "German honor." The sailors refused to get up steam. The officers attempted to discipline the mutineers. The mutineers broke into arms lockers and armories and took to the streets. The U-boat sailors joined their comrades from the surface ships. It was all over. All German naval operations had ended.

But twenty-one years later, the U-boats would return, many more U-boats, faster, more powerful U-boats. And the war they waged would be even deadlier.

The Battle of the Atlantic

"The only thing that really frightened me during the war was the U-boat peril," Winston Churchill said. "It did not take the form of flaring battles and glittering achievements, it manifested itself through statistics, diagrams and curves unknown to the nation, incomprehensible to the public."[10]

At the beginning of World War II, U-boats were far from Adolf Hitler's top priority. In 1939, German admiral Karl Dönitz, chief of the submarine division, had fewer subs than the United States—57 as opposed to 112. And of those fifty-seven U-boats, Dönitz had only twenty-seven with enough range to use on the ocean. Dönitz told Hitler he'd need at least 300 oceangoing subs to properly blockàde Britain. Hitler expected the war to be over long before Germany could build even a third of that many U-boats.

In World War I, when detection devices ranged from primitive to nonexistent, Germany had been able to sneak submarines through the English Channel. This was no longer possible. Now, to get to the Atlantic a German submarine would have to sail north around the tip of Scotland. Dönitz had only eight subs with a range of 12,000 miles, making them truly oceanic. Eighteen more could sail as far as Gibraltar and back, about 6,000 miles. The rest were confined to the North Sea.[11]

Everything changed after the fall of France. Britain didn't sue for peace, as Hitler expected. The Luftwaffe failed to defeat the Royal Air Force (RAF) and it hardly touched the Royal Navy, so an invasion of Britain had to be indefinitely postponed. Now, it became necessary to wage an all-out campaign against British shipping. What's more, the fall of France made it easier to do so. The German navy now had the entire Atlantic coast of France from which to send U-boats. The British could certainly close the Strait of Dover, but it couldn't block the Bay of Biscay so easily. U-boats sailing out of the French ports were right on the main shipping route to the Mediterranean, Africa, and Argentina, a major supplier of beef and wheat. They concentrated on that sea-lane and sank ships in the eastern Atlantic. Occasionally, one of the longer range subs would go into the Mediterranean.

The British did not revive the convoy system immediately. There were not enough destroyers, frigates, and corvettes to act as escorts. In the tight economic environment of the 1930s, it was easier to get appropriations for battleships and cruisers than for smaller vessels. When convoys were revived, they proved to be less effective than they were in the last war.

For one thing, antisubmarine equipment had not improved nearly as much as submarines. The British had a sound-ranging system called Asdic—the same as the American sonar—but it could locate a sub only if the U-boat was within 1,000 yards. Also, the early Asdic systems could not determine the sub's depth. Depth charges are set to explode at a predetermined depth. So depth charges to be used against U-boats would have to have their depth fuses set by guess. The U-boats, on the other hand, were much bigger and sturdier than those of World War I. Depth charges had to explode close to them to do damage, and they usually could not be put out of action by a single hit in a surface fight. The new subs were also much faster, especially on the surface, than those of World War I.

This surface speed made it possible for Dönitz to develop "wolf pack" tactics. U-boats would be scattered along the sea-lanes. If one of them spotted a convoy, it would notify headquarters, which would then order twenty or more U-boats to converge on the convoy and then coordinate the attack. That was far more submarines than the naval escort could handle.

The United States had been moving from neutrality to increasing involvement in the war. American shipyards began working overtime to replace ships lost in the Battle of the Atlantic. The Americans would launch 1,500 new ships a month—three times more than the Germans could sink—and they would be large freighters, many of them of 10,000 or 15,000 tons. Replacing lost ships might be considered passive resistance to the U-boat offensive. However, active resistance was developing at the same time. In addition to building new freighters, the U.S. shipyards were also building up American naval strength—more ships for the navy that in 1939 was already the largest in the world. Meanwhile, the United States, under its "lend-lease" program, transferred fifty old destroyers to the Royal Navy.

Next, the United States declared that the western North Atlantic was off-limits for U-boats. At first, it merely informed the British of the presence of U-boats it found. Later, it began to sink them. U.S. and British air bases now allowed planes to patrol the entire North Atlantic, which was becoming a rather unhealthy place for German submarines. U-boat commanders probably celebrated when Japan bombed Pearl Harbor and the United States entered World War II. They switched their operations from the increasingly hostile North Atlantic to the Atlantic and Gulf coasts of the United States.

Most of the U.S. Navy ships suitable for convoy work were on duty in the North Atlantic. At first, the huge U.S. Navy did not have enough escort ships to organize convoys along the coasts of North America. Much of the blame for this situation belongs to Admiral Ernest J. King, chief of naval operations, who seemed to be expecting a Jutland-style battleship battle in the North Atlantic. In late 1941 and early 1942, King had collected the navy's most modern battleships and aircraft carriers and stationed them on the East Coast, where they could do little about the submarine menace and were not likely to have a showdown with the weak and bottled-up German surface fleet. At the same time, the capital ships were away from the Pacific, where they would be most helpful in countering the formidable Japanese navy. While building up an Atlantic fleet of big warships, King neglected to ask for enough destroyers, destroyer escorts, sub chasers, and other small vessels suitable for convoy duty.

The U-boat crews referred to this period of operations in North American coastal waters as the "happy time." Sinkings by U-boats skyrocketed. Convoys were eventually organized, and the U-boats gradually shifted their operations farther south—to the Caribbean and the coast of South America.

Apart from the shortage of escort craft, there was one problem that had been present since the beginning of the war. In a way, both the British and U.S. navies were flying blind.

The Intelligence War

Both the British and the Germans intercepted each other's radio messages and avidly tried to learn the other side's moves in the Battle of the Atlantic. The British rerouted their convoys from areas they considered dangerous, and the Germans wanted to learn the new routes. At the same time, the British wanted to know the location of the U-boats and the ships that supplied them. At first, the Germans had an advantage. The Royal Navy, unlike the Royal Army and the RAF, used an antiquated cipher system, based on code books and the use of pencil and paper. Although the ciphers changed regularly, the German navy's Observation Service was able to read much of the British naval communications. Sometimes, the Germans got information on a convoy's route ten to twenty hours in advance.[12] The result was a disaster for the convoys.

British decoders were at first less successful, even though they knew far more about the German codes than the Germans suspected.

German messages were encoded by the Enigma, a machine that scrambled characters so thoroughly the Germans believed they were completely unreadable. But Hans-Thilo Schmidt, an embittered German, sold information about Enigma to the French in 1931 and continued to update it until the outbreak of the war. French experts looked at the information and decided that the Germans were right: Enigma codes were unbreakable.

The Poles, however, did not think Enigma was impregnable. They had already decoded commercial messages sent through commercial Enigma machines, but so far the German government Enigma machines had foiled them. The Polish Encryption Bureau gave the task of solving the Enigma to Marian Rejewski, a young Polish mathematician. He solved it. For some time, the Poles were able to read all German government messages. Then the Germans added two more scramblers to the three in the original machine, which enormously increased the possible code keys.

The simplest of ciphers is a substitution system. *The Lone Ranger*, a radio show for children, had a cipher for the young fans who joined the Lone Ranger Club. It simply substituted the letter B for A and so on down the alphabet. B in "plaintext" was C, C was D, down to Z, which in *Lone Ranger* code was A. Simple ciphers like this have been broken for thousands of years by "frequency analysis"—E is the most frequently used letter in the English language, followed by T, and so on. The rules for frequency analysis appear in Edgar Allen Poe's story *The Gold Bug*. The Enigma used a substitution system, but instead of a single "encrypted alphabet," like the *Lone Ranger* code, the scramblers in the machine put each letter into a different encrypted alphabet. And there were three scramblers in the original Enigma. So the number of encrypted alphabets equaled $26 \times 26 \times 26$, or 17,576. If the scramblers had only one setting, there would be that many possibilities to be tested. But there were 17,576 ways to set the scramblers. These settings, with settings to another device called a plug board, were called keys to the Enigma code. Then, just before the invasion of Poland, the Germans added two more scramblers to the machine and more cables to the plug board. The result was that there were now

159,000,000,000,000,000,000,000 possible keys to an Enigma message. And the Enigma operators changed the keys frequently during the day. These additions stymied Rejewski, who had been building computer-like machines called "bombes" to unscramble the scramblers. He ran out of money and time, and the Germans overran Poland, which was unable to decipher their signals. Just before the attack, Rejewski's superior officer telephoned the heads of the French and British encryption bureaus and gave them information about all the work his bureau had done on the Enigma. The French, who believed the German machine was impregnable, were astounded.

The British set up a new code breaking organization. One of their code breakers was Alan Turing, who designed a kind of computer and picked up where Rejewski left off. The British began deciphering more and more German messages. Soon, the only area where they lacked any significant success was on the German naval code. There was a reason: while the other German government agencies used an Enigma machine with five scramblers, the navy used one with eight scramblers, which boosted the possible Enigma keys to an astronomical number. It was time to try another approach: theft.

Ian Fleming, the creator of James Bond, was a member of British Naval Intelligence at that time. He proposed crashing a captured German bomber near a German ship. The crew of the bomber, German-speaking British agents, would pose as Luftwaffe members and be rescued. Then, they would steal the Enigma documents showing the starting key for each day. From there, Turing and his fellow code breakers would have a head start. The operation, code-named Ruthless, was about to begin when the naval chiefs noticed that there was no German shipping conveniently located. Operation Ruthless had to be postponed indefinitely.

Four days after the postponement, Frank Birch, head of the naval section of Bletchley Park, the British code-breaking center, said Turing and a colleague, Peter Twinn, came to see him.

"Turing and Twinn came to me like undertakers cheated of a nice corpse two days ago," Birch said, "all in a stew about the cancellation of Operation Ruthless."[13]

This postponement didn't stop attempts to snatch information on the daily keys, however. There were other raids, several of them—and, unlike the movie, conducted by British, not Americans. They recovered enough information to enable the Bletchley Park cryptanalysts to decipher most of the German naval messages and set the stage for what a German writer called . . .

"1943—The Year of the Slaughter of the U-Boats"

Every ship the British seized documents from was sunk, so Berlin would think that the daily keys had gone to the bottom. They were also careful not to sink every U-boat or supply ship they learned about, because they didn't want the Germans to think they had the Enigma documents. On one occasion, they learned of the location of nine German tankers and supply ships. The code breakers gave the Royal Navy the locations of only seven of them. Unfortunately, British destroyers stumbled on the two remaining ships that were supposed to have escaped and sank them. For a moment, the German naval authorities were afraid their code may have been broken. But on reflection, they decided that since Enigma messages were impregnable, the clean sweep was just a bad break

Breaking the code was a giant step toward eliminating the U-boat menace, but there were other elements that made 1943 a turning point. Detection of submarines had become easier. Sonar (or Asdic, as the British called it) had greatly improved: depth-charge fuses could now be set more precisely. Radar had improved immensely. It could now be carried aboard planes as well as ships. And there were more planes looking for U-boats. Besides patrol planes, the U.S. Navy had a fleet of blimps, which were slow compared to airplanes but faster than any naval ship and far faster than any U-boat, able to hover over one spot, and with more than twice the range of any airplane. Most important, there were now escort carriers—small carriers that could provide air cover for the convoys.

The planes carried bombs, depth charges, rockets, and automatic cannons. One version of the B-25 carried a 75 mm gun. Probably the most formidable weapon the planes carried was the homing torpedo. The torpedo automatically steered for the sound of the sub's propellers. Surface ships carried homing torpedoes, too, and they also had new depth-charge launchers like the hedgehog, which shot a pattern of 31 shells 250 yards ahead of the ship.

In 1943, all of these items went into service. The Allies knew the position of the German submarines and were able to divert their convoys at the last minute and at the same time attack the subs and their supply ships. In mid-1943, sinkings of commercial shipping dropped precipitously, while the sinking of U-boats increased proportionately. On May 24, 1943, Dönitz withdrew his submarines from the Atlantic, saying, "We have lost the Battle of the Atlantic."[14]

They lost heavily. Germany sent out 830 U-boats. Of them, 696 were sunk. Among U-boat personnel, the death rate was 63 percent, while the overall casualty rate was 75 percent killed, wounded, or captured. This was the highest casualty rate of any service of any country in the war.

Regardless, the Germans continued to improve their submarines. In 1944, they put out subs equipped with *schnorkels* (snorkels)—a breathing device that projected above the surface like a periscope and let the sub use its diesel engines while submerged. Schnorkels made it harder for radar to pick up a submarine, but they came too late. Hydrogen peroxide engines were added in 1945. They let the U-boat use an internal combustion engine fully submerged. But by that time, it was far too late.

After leaving the North Atlantic, the U-boats tried to operate in the South Atlantic and even the Indian Ocean. Targets were scarcer in those areas, though, and the Allies, particularly the United States, began sending out "hunter-killer" teams composed of an escort carrier and several destroyers or destroyer-escorts. Some of their encounters were reminiscent of the battles of John Paul Jones.

On November 1, 1942, the destroyer *Borie* located a sub and dropped depth charges. The sub surfaced, and *Borie* fired at it with its four-inch deck gun. The sub fired back. *Borie* tried to ram the sub, which was not a good idea, because *Borie*, an ancient four-stacker, had plates that were rusted paper-thin over the years. It sprang leaks, but its momentum carried it right over the deck of the U-boat. The crews of the two vessels began firing at each other with rifles, SMGs, and pistols. One sailor even threw a sheath knife and hit a German trying to man the sub's deck gun. After ten minutes, the submarine broke away, then turned and attempted to ram *Borie*. *Borie* turned and launched depth charges over its stern. Three depth charges exploded under the pursuing submarine, lifting it out of the water. The sub turned again and attempted to flee. *Borie* pursued and opened fire. One shell blew the Germans on the sub's bridge overboard. After a second hit, the U-boat surrendered.[15]

On the other side of the world, the submarine war was entirely different.

1943—The Year of Slaughter *by* the Submarines

It would seem that the United States, with forces operating at the far end of the Pacific and enormously long supply lines, would be especially vulnerable to submarines. Furthermore, the Japanese subs carried the "long lance," undoubtedly the best torpedo of World War II. But the Japanese did relatively little damage to the ships that supplied U.S. forces in the Pacific. It seems that Japanese sub commanders considered attacking supply ships unworthy of a Samurai. Honor could only be gained by attacking warships. The trouble was that warships were tough and dangerous targets. Early in the war, the Japanese subs were somewhat successful in attacking surface warships, but American antisubmarine efforts changed that, particularly as Americans gained control of the air over the Pacific. At the Battle of Leyte Gulf, the Japanese had sixteen submarines, but they managed to sink only one destroyer escort.[16] At the opening of that battle, two American submarines, *Darter* and *Dace*, sank two Japanese battleships, including the Japanese flagship, and disabled a third.[17] Bombarding enemy shore positions was also honorable—honorable but mostly useless. Japanese submarines cruised thousands of miles on these absurd operations.

When American naval power made travel between the Pacific islands hazardous, Japanese submarines were increasingly pulled off patrol to transport troops and supplies. Before long, there were so few Japanese submarines threatening supply lines that the United States took escort vessels off convoy duty and assigned them to fleet actions.

At the beginning of the war, American submarines were also intended to attack warships. But U.S. military planners quickly saw that an island nation like Japan was vulnerable in the same way Britain was. The U.S. subs began attacking Japanese commercial shipping as well as warships. U.S. submarine strategy was different from the German. American submarines were generally bigger than the German U-boats and had a longer range. They didn't need supply ships and there were no wolf packs and little radio communications. Japanese antisubmarine tactics were primitive, and U.S. captains were expected to act on their own initiative.

For the first year, they weren't very effective, though. Because of the economy in the prewar years, torpedo maintenance had been shortchanged. Torpedoes missed their targets and failed to explode when they hit. In 1943, they were updated, which marked the beginning of the end of Japanese shipping. In 1943, American submarines sank 22 Japanese warships and 296 merchant ships. In 1944, the submarine attack picked up steam. In the mammoth assault on Truk, planes from the carrier task force sank five Japanese tankers. Then the submarine U.S.S. *Jack* attacked a convoy of five tankers escorted by three destroyers. *Jack* sank four of the tankers by itself. On November 29, 1944, the U.S.S. *Archerfish* made the biggest submarine kill of the war: the new 59,000-ton carrier *Sinano*, on its maiden voyage. Less than a month later, the U.S.S. *Redfish* hit another carrier, the 24,000-ton *Junyo*, and knocked it out of the war. Ten days later, on December 19, *Redfish* sank another carrier, the new *Unryu*, even though it was escorted by three destroyers. By war's end, U.S. submarines sank 2,117 Japanese merchant ships, totalling 8 million tons. Japan had only 1.8 million tons of commercial shipping left—mostly tiny vessels on the Inland Sea. Submarines sank 60 percent of all Japanese commercial ships destroyed; aircraft, 30 percent; surface ships and mines,

10 percent. And the subs sank 201 of the 686 Japanese warships sunk during the war.[18]

All this with seventy-five boats—a record the vaunted German U-boat fleet couldn't match.

The submarine had caused a major turning point in naval warfare, first in World War I, then in World War II. It is poised to create an even greater turning point in the future, but we'll look at that in chapter 20.

17

Blitzkrieg and Antiblitz

Troops of the 55th Armored Infantry Battalion and a tank of the 22nd Tank Battalion advance in Wernberg, Germany in 1945. The Allies adopted Blitzkrieg tactics and used them against the Germans. (U.S. Army photo, National Archives.)

The Politics of Revenge

IN THE 1930s, the German army demonstrated the advantages of losing a war. The losing army was not burdened with a lot of obsolete or obsolescent weapons, nor did it have to base its tactics on these weapons. And, as its old methods had so obviously failed, it was encouraged to look for new ones.

At the end of World War I, that bloodbath excited a common attitude about war among all the people of Europe: revulsion. Among the professional military of the victors, the most common feeling was complacency, and among the victors' political leaders, the need to economize. Among the leadership of the losers, however, the predominant feeling was a desire for revenge, which was particularly true for Germany. The Treaty of Versailles gave Germany's colonies to the victors. Redistribution of colonies was the usual result of a European war, as was the rearrangement of the loser's borders. But the Treaty of Versailles went further. It limited the Germans' navy to little more than a coast guard and their army to a token force. Also, Germany wasn't allowed to build or maintain an air force, submarines, battleships, machine guns, submachine guns, heavy artillery, or tanks. Nor could there be compulsory military service. And Germany couldn't join the newly formed League of Nations. All this was based on the theory that the Germans loved to make war on their neighbors. In 1914, Germany was no more enthusiastic about war than France, Russia, or Austria-Hungary, but Versailles changed that. In the postwar period, Germans, racked by runaway inflation and civil strife, turned to leaders who blamed all their troubles on the victorious Allies and promised revenge. The Germans elected Adolf Hitler.

Their actions showed once again what thousands of years of history have proved: revenge makes poor foreign policy. Fortunately, Henry Morgenthau's plan to reduce Germany to a cow pasture after World War II was rejected and the Marshall Plan was instituted.

The Germans began to rearm. At first, it was done secretly. The government organized a huge glider program to train future Luftwaffe pilots. It started the Lufthansa commercial airline for the same reason. Karl Dönitz developed his wolf pack submarine tactics using motor boats. German small arms experts developed the "general purpose" machine gun—the model for all modern machine guns—and had several types manufactured in Switzerland. Germany worked out an arrangement with the Soviet Union to manufacture tanks and train soldiers in their use there. Later, when the Allies indicated that they didn't care about German rearmament, Hitler became more open. This was especially true after the French failed to prevent his remilitarization of the Rhineland.

Since the Germans had no old weapons, except a few Mauser rifles used by their tiny army, they could start from scratch. They didn't have any old, useless hardware to discard. However, they did have some old, useless ideas, ideas that the late war had proved to be a handicap.

New Ideas for a New Army

One of the first old ideas to go was the old master-slave relationship between officers and enlisted men—a feature of the old Prussian army that was still characteristic of the French army and, to a lesser extent, the British. This and other changes were brought about by Hans von Seeckt, a round-faced, monocle-wearing Prussian who was another of the remarkably brainy generals to be found on Germany's eastern front in the Great War.[1] Seeckt had been August von Mackensen's chief of staff and became commander in chief of the German army under the Weimar Republic, which succeeded the German empire. His guiding principle was to "neutralize the poison" of the Versailles Treaty. Seeckt's new army was composed of all volunteers, each man carefully vetted. The new, comradely relationship between officers and men fostered mutual confidence and, it was hoped, would prevent the sort of mutinies experienced by almost all belligerents in World War I. And those enlisted men were destined to play an important part in Germany's military future. In the postwar army of 100,000, there were 40,000 noncommissioned officers (NCOs)—all of whom would be officers when the army expanded. As the NCOs were trained for higher ranks, so were the officers. Each platoon leader was trained to command a battalion, and each major to lead a division.

The new German army discarded most of the old trench-fighting tactics of the last war. In the new army, everyone was trained to be a storm trooper.[2] German industry developed weapons to suit storm trooper tactics. Troops attacking in small groups, finding their way between strong points, sometimes isolated in the enemy's rear, needed a lot of firepower. In 1918, they used the Maxim 08/15—a reliable gun, but the heaviest "light machine gun" in the war, which therefore handicapped mobility. They also had a sprinkling of submachine guns—light weapons that lacked power, range, and accuracy. The Allies had light machine guns. The Lewis gun and the Benet-Mercie were lighter than the Maxim, but they jammed too often. The Browning automatic rifle was even lighter than the other two and quite reliable, but it lacked fire capacity. It was fed by detachable twenty-round magazines, and if fired too fast, the barrel became red-hot and useless. German leaders wanted something better.

What they got was the general purpose (GP) machine gun. The GP gun was about the same weight as the Lewis, but it was absolutely reliable. Unlike the Lewis, it could be fired almost continuously. If a barrel became overheated, it took only three or four seconds to remove it and replace it with a fresh barrel. The GP gun could be fired from a bipod like a light machine gun or mounted on a tripod for sustained or antiaircraft fire. It could even fire over the heads of friendly troops to harass enemy positions with indirect fire.

During the secret rearmament period, parts for these machine guns were made in German and Austrian factories and assembled in a German-owned factory in Switzerland. The same system was used to produce submachine guns (SMGs) for the projected storm trooper army. The SMG provided short-range firepower and was issued much more widely than in the Allied armies. For years, the British and U.S. armies disdained the SMG as a "gangster gun." After Germany began rearming openly, it officially adopted a new SMG, the MP (for *Maschinenpistole*) 38, a metal and plastic design that became a model for both the British Sten and American M 3 SMGs, as well as for most other modern SMGs.[3]

Mechanizing the New Army

In the last war, the Germans made little use of the tank. That, Seeckt believed, was a mistake. So did his successors, when the founder of the new army retired in 1926. They began working out ways to incorporate tanks into the new storm trooper tactics. They also thought there was a place for that other new, and greatly underemployed, machine: the airplane.

A number of military thinkers in Germany's former enemies had been working on ways to employ tanks. In the Soviet Union, the strange Marshal Mikhail Tukhachevsky, who once said of the Russian army, "It is a horde, and its strength is the strength of a horde," also favored masses of tanks.[4] And the Soviet Union built masses of tanks. Tukhachevsky advocated having concentrations of artillery, tanks, and infantry break through the enemy line, then the tanks would pour through the gap, separate enemy formations, and encircle them so they could be annihilated by artillery and armored forces. During the period the Soviets were building their armored force, they purchased high-speed tanks designed by the American J. Walter Christie, which had eight large wheels and could run with or without tracks.[5] Development of Christie's tank culminated in the Red Army's superb T-34, universally conceded to be the best tank of World War II. Red Army maneuvers in 1936 greatly impressed General Lucien Loizeau, the head of the French military delegation, who wrote, "The technique of the Red Army is on a particularly high level . . . To achieve this level of armaments in three or four years not only demonstrates the power of Soviet industry, it also establishes the immense superiority of the Red Army over all other European armies, which are often forced, and for a long time, to use old material."[6]

But in 1937 Joseph Stalin had Tukhachevsky shot during his purge of the military, and for some time the Soviet Union lacked a doctrine on how to use all those tanks.[7]

In France, the economy and the opinions of old-school generals set back progress in armored warfare in a less bloody fashion. France had heavily invested in the Renault light tank during the war and could not afford to junk that machine. The enormous expense of the Maginot Line caused other military spending to be strictly limited. The Renault tank was slow, lightly armored, and meant to accompany the infantry. France later acquired better tanks, some better than the best German machines, but it continued to use them as "roving pillboxes" among the infantry. Colonel Charles de Gaulle advocated tactics using tanks as the main element instead of as merely support for the infantry, but the military and political hierarchy ignored him.

The U.S. Army had a similar problem. Almost all of the thousand French Renaults and the hundred British Mark V tanks it had acquired in the Great War were still in good working order. Nevertheless, in 1930 the army's high command authorized the creation of a permanent tank corps, separate from any other branch. That lasted until Douglas MacArthur became chief of staff. The separate tank corps was disbanded in 1931.

Britain for a while led the world in the development of tank tactics. General J. F. C. Fuller, an armored warfare enthusiast and an indefatigable propagandist, got the British army to establish the Royal Tank Corps. His work attracted the attention of Captain Basil H. Liddell Hart, a soldier who had been invalided out of the army because of injuries from poison gas and who then became a military historian and commentator. Fuller and Liddell Hart believed tanks must play an independent role, rather than being tied to the infantry.

Fuller favored using fast medium tanks to sweep around both enemy flanks while heavy tanks and motorized infantry attacked frontally. Liddell Hart preached the doctrine of the "expanding torrent" of tanks. These would break through the enemy lines using a mechanized version of the German storm troopers' infiltration tactics. The commander would pour reserve tanks into the breaches, and the "expanding torrent" would destroy enemy communications and command posts and attack the enemy from the rear. This "psychological dislocation," Liddell Hart said, would bring victory quicker— though just as complete—as the orthodox objective of annihilating enemy forces.

The writings of Fuller and Liddell Hart gained a number of military converts to the cause of armored warfare. Eventually, British army authorities authorized an experimental mechanized brigade to test the theories. The operations of this unit attracted European and American military observers. In 1929, the British army completed the report *Mechanized and Armoured Formations*. The report was supposed to have restricted circulation, but it turned up in Germany, where it was read with interest by Colonel Heinz Guderian, Germany's leading armor expert. But the worldwide Great Depression put an end to the experiments.

Guderian began training a tank force before Germany had any tanks by using trucks. He had never even seen a tank before 1929, when he saw one in Sweden. Later, with the German-Soviet agreement, German troops began using real tanks on Soviet territory. Finally, with Hitler firmly in power, Germany began building its own tanks. At first, these machines, like all the rest of the world's tanks, were undergunned. But as early as 1935, Guderian began demanding a tank with a 75 mm gun, and work began on what was to become the Pzkw (for *Panzerkampfwagen*—armored fighting vehicle) IV, Germany's main battle tank in World War II.

Guderian was deeply influenced by the writings of Fuller and Liddell Hart. He developed their mechanized brigade into the panzer division, then the panzer corps by adding a mechanized infantry division. When the Battle of France began, the German attack was spearheaded by what amounted to a panzer army. Several panzer corps broke through Allied lines just north of the Maginot Line and swept up to the English Channel, trapping all of the British forces and part of the French army. Most of those troops were evacuated at Dunkirk, and the Germans turned to the south and west to deal with the rest of the French army.

Components of the Blitzkrieg

The speed of Germany's victory in Poland had been surprising, but no one thought the Polish army was a match for the German army. And besides, the Poles had been stabbed in the back by Germany's then-ally, the Soviet Union. But the fall of France was shocking. The French were generally believed to have the best army in the world. The Germans had utterly trounced the combined armies of Britain and France in a month and a half. The news media coined a new word for Germany's way of war: blitzkrieg (lightning war).

The blitzkrieg was based on masses of tanks, followed by infantry in armored personnel carriers (panzer grenadiers) and self-propelled artillery. It used infiltration tactics to break through enemy lines and attacked rear areas, disrupting enemy communications and command structures. The panzers invariably tried to encircle enemy forces.

This was the ideal. In actuality, the German Wehrmacht never had enough armored personnel carriers and it was seriously short of self-propelled artillery—a shortage that lasted through the whole war. It was successful, however, because its enemies had dispersed their tanks instead of massing them, because enemy troops were seriously shaken when the panzers got in their rear, and because the Germans had developed a reasonable substitute for artillery: air power.

Although Germany, like Britain, had an independent air force, the Luftwaffe's orientation was entirely different from that of the Royal Air Force (RAF). The Luftwaffe's purpose was primarily to support the ground troops, while the RAF's was strategic bombing. Army and Luftwaffe officers were routinely transferred from one service to another.

Germany always attacked without warning. Its first move was to destroy as many enemy aircraft on the ground as possible. This task was usually given to the *Stukas,* the dive bombers, especially the Junkers JU 87. Dive bombing was a technique that had first been developed by the U.S. Navy and was copied by other navies. It appealed to navies, because it was the most accurate way for a plane to deliver a bomb, and naval bombers had a greater need for accuracy than most others. Their target was a ship—a very small object from the air—that might be moving more than forty miles per hour and could turn while the bomb from a typical land-based bomber was falling. The Germans adopted the dive bomber for land warfare, because they needed a substitute for the artillery that was often unable to keep up with the panzers and the motorized infantry. Only dive bombers could deliver explosives with something like the accuracy of artillery.

As an airplane, the Stuka was inferior in range, speed, and armament to such U.S. naval dive bombers as the Curtis Helldiver or the Douglas Dauntless. It was a new and terrifying weapon to most European troops, though. The plane itself, with its fixed landing gear, looked like a diving hawk, and sirens attached to it emitted a frightening scream as it dived. Dive bombers, because they presented an almost unmoving image to a gunner below them, were extremely dangerous to operate if the enemy gunner was not panicked. But the Stukas managed to panic a lot of troops, and they were accurate. They were used at the front to eliminate pill boxes, bunkers, and gun emplacements and on rear areas to destroy or cripple railroads, marshaling yards, bridges, and distribution centers. Farther back, German medium bombers hit supply lines to inhibit enemy attempts to reinforce threatened areas. German heavy bombers attacked factories, warehouses, and cities.

To seize bridges, reduce fortresses, and perform other tasks to smooth the way for the panzers, the Germans used airborne troops. Both the Germans and the Russians had been training parachute troops for several years before the war, but paratroops made their first combat appearance in Holland. In Belgium, the Germans used another kind of airborne force: glider troops. They landed glider troops on the top of Fort Eben Emael, once considered the strongest fortress in Europe. The fort's guns commanded every inch of the terrain around it, but none of them pointed skyward. Nine gliders landed on top of the fort, and German troopers used shaped demolition charges to blow holes in the turrets and bunkers and flamethrowers to attack the defenders. The Belgians barricaded the tunnels of the fort, and the battle raged underground until the next day when the panzers and infantry arrived.[8]

The blitzkrieg was monstrously successful not only because of its tactics, but also because of the strategy that employed those tactics. The strategy was to encircle enemy forces and destroy them if they did not surrender. By previously annexing Czechoslovakia, it had a base from which to attack Poland from the south, while other German troops moved down from East Prussia. Germany moved troops into East Prussia by sea and also by land, encircling several Polish divisions on the way. Then the troops in East Prussia pushed south to meet those from Czechoslovakia. Polish troops who were not encircled tried to retreat to the Pripet Marshes—difficult terrain for panzers—but were attacked from the rear by the Red Army.

In France, the French and British field armies concentrated on the

Belgian frontier. They believed the Maginot Line secured the southern part of the border and that the Germans would revive the Schlieffen Plan. They were almost right. The German General Staff did want to execute an attack similar to the Schlieffen Plan (although this time they'd invade the Netherlands as well as Belgium and Luxemburg). Hitler himself wanted to break through just north of the Maginot Line, but most of the generals said the Ardennes forest and hills were not suitable for panzers. The former lance corporal (equivalent of an American private first class) bowed to the superior military wisdom of the generals. But then a German plane carrying elements of the plan crashed in Belgium. Hitler turned to the generals who objected to the new Schlieffen Plan: Gerd von Runstedt and his chief of staff, Erich von Manstein. Manstein consulted Guderian on the feasibility of a panzer attack in the Ardennes. Guderian said it would work if there were enough tanks. So the blitzkrieg broke through the Allied line just north of Sedan and encircled all the troops between Sedan and the English Channel.

Downfall of the Blitzkrieg

A major contribution to the success of the blitzkrieg was the ineptness of its opponents. In Germany's attack on the Soviet Union, the Soviets showed massive ineptness and suffered massive losses in the first few months. German panzers encircled huge numbers of Russian troops, while the Soviet soldiers followed Stalin's orders to "fight to the last bullet." To a large extent, this order was responsible for the Soviet Union's staggering loss of 7 million soldiers killed in combat. Another reason was the way Soviet generals lavishly expended manpower when on the offensive. Sending tanks and infantry over known minefields is one example. Another is the Soviet phenomenon of "tank riders," infantrymen perched on moving tanks who were expected to use hand grenades and submachine guns against enemy infantry. A single burst from a German machine gun could kill half a dozen of those exposed tank riders.

Russian weather—the mud, then the cold—stalled the German offensive, and the Soviets were able to profit from the hard lessons they had learned at the beginning of the war. They attacked the Germans with masses of tanks, and in the T-34, they had a better tank. They also practiced encirclement, especially at Stalingrad and at Kursk. With more or less evenly matched tank forces, the blitz had gone out of the blitzkrieg.

It was not likely to return. Besides the universal appreciation of how to use armor, a horde of new antitank weapons had appeared. When the war began, the primary antitank gun in all armies was a small, low-lying, high-velocity gun. Those early guns were handy—they weighed less than a ton and were easy to manhandle and hide. They'd furnish the infantry protection against the armored monsters. If only they could stop tanks. The caliber of the guns was between 37 and 47 mm—roughly from 1.5 to 2 inches in diameter. A rule of thumb holds that the penetration of an artillery shot is equal to its caliber. High velocity improves penetration.[9] But tank armor quickly became too thick for the little cannons to penetrate. Antitank guns had to get bigger and bigger. The Germans ended up with a monster of 126 mm caliber (about 5 inches). Guns like that are far too big for infantry. The armies made their antitank guns self-propelled and called them "tank destroyers." They were no longer infantry weapons.

But infantry had other weapons.

Early in the war, the best way to penetrate tank armor was with solid shot—a projectile that looked like an artillery shell but was solid metal. But guns firing three-inch or larger shot were too heavy for infantry to maneuver. So the designers turned to shaped charges. It had been known for years that an explosion with a cone-shaped depression in it would blast a jet of flame out of the cone and burn through steel like a cutting torch. It was most effective when detonated a certain distance (depending on the size of the cone and the weight of the charge) from the metal to be penetrated. This peculiar effect was discovered in the 1880s by an American engineer named Monroe, and is often called the Monroe effect. In the 1920s, a German named Neumann learned that lining the cavity with metal created a deeper hole in the target. The Germans used shaped charges to blast into the gun emplacements at Eben Emael, and all countries began producing artillery shells with shaped charges. There was a drawback to using shaped charges in artillery shells, though. The gun's rifling caused the shell to spin, which reduced the power of the burning jet. In 1942, the United States introduced the bazooka, a 2.36-inch rocket launcher using a shaped charge. The bazooka was a light weapon and could be fired from an infantryman's shoulder. It was later improved to 3.5-inch caliber and dubbed the super bazooka. It fired a fin-stabilized rocket that did not spin. The same principle made rockets fired from airplanes, and later helicopters, most deadly antitank weapons.

The British had the PIAT (Projector Infantry Anti-Tank), a strange contraption that used a heavy spring and a charge of powder to shoot an antitank shell. They soon traded it in for the bazooka. The Soviets had one of the simplest antitank weapons: a hand grenade that trailed a streamer so it would land point first. It was called the RPG 43.[10] Incidentally, the letters RPG do not stand for "rocket propelled grenade." They are used on all one-man-portable Russian antitank weapons like the postwar RPG 2, a recoilless gun, and the RPG 7, a recoilless gun cum rocket, widely used in Vietnam and Iraq. The Germans and Japanese also had antitank hand grenades. The Japanese, too, had the "lunge mine," a weapon that's hard to imagine any nation but the suicidally inclined Japanese adopting. It was a shaped charge on the end of a long pole. The soldier would get as close to the tank as he could, then he'd ram the lunge mine into it. If the tank didn't get him, his own mine would.

Perhaps the most feared antitank weapon of World War II was the German *panzerfaust*—feared by both enemy tankers and panzerfaust gunners. The panzerfaust was a type of recoilless gun. A recoilless gun is basically a tube with two openings. The projectile goes through the large opening, the muzzle. Through the small opening goes gas at very high speed. Both are the result of the ignition of the powder charge. The gas, ejected from a venturi at the rear of the weapon, has enough velocity, and therefore enough energy, to balance the muzzle energy of the projectile, so there is no recoil. Standard recoilless guns, in calibers from 57 mm to 106 mm, are useful for antitank work, especially the larger calibers, which have a jet blast that's lethal in spite of the spin. But the panzerfaust, because it was so small and light and had no spin, was in a class by itself. In the panzerfaust, most of the projectile sat on the end of the tube. The projectile contained a heavy enough shaped charge of explosive to penetrate any tank, but it wasn't very accurate. The gunner had to get close to his target. That fact, and the chance that the thing might blow up in its user's hands, made the panzerfaust one of the Wehrmacht's least popular weapons. The Soviets saw that it had possibilities, though. They improved it and produced the RPG 2, which had a longer range. Next, they added a rocket motor to the projectile and called it the RPG 7—a powerful, relatively long-range infantry weapon that has been featured in Vietnam and in about every conflict since.

Antitank warfare also led to an enormous increase in the use of land mines in the many little wars—as well as the two big wars, Korea and Vietnam—that have plagued the world since 1945. Land mines have been even more ubiquitous than the RPG 7.

By demonstrating what tanks and close air support could do, the blitzkrieg introduced a major turning point in war. The reaction to the blitzkrieg, with its proliferation of antitank weapons, also changed war. And the huge number of antitank weapons that can be used by individuals has vastly increased the potency of guerrillas.

18

The New Queen of the Seas

U.S. ships fill the sky with antiaircraft fire during the battle of Santa Cruz, October 26, 1942. Barely visible is the splash of a bomb that has just landed behind the carrier in the foreground, dropped by the Japanese directly above it. (U.S. Navy photo, National Archives.)

Tora! Tora! Tora!

COMMANDER FUCHIDA MITSUO peered intently at the array of ships moored in the harbor. There was no sign of activity. He spoke into his microphone and gave the signal all his shipmates had been waiting for, "Tora! Tora! Tora! (Tiger! Tiger! Tiger!)," meaning the surprise

221

would be complete. He fired a signal flare. His dive bombers dived at the motionless fleet.

On the minelayer *Oglala*, this early Sunday morning, Rear Admiral William R. Furlong noticed a plane drop a bomb near the edge of the harbor. Furlong thought it was an accident, then he noticed the red circle—Japan's Rising Sun symbol—on the plane's wings.

"Japanese! Man your stations!" Furlong yelled. *Oglala* signaled, "All ships in harbor sortie."[1] The ships, however, didn't have time to start their engines. Enemy dive bombers were dropping bombs everywhere, but especially at "Battleship Row." Before the American sailors could get their antiaircraft guns into action, the Japanese torpedo bombers swept over the roof tops of Pearl City, north of the harbor, and over the naval station, south of the harbor.

Fuchida had wanted the torpedo planes, his most potent weapons, to lead the attack, but the faster dive bombers seemed to be too eager. Actually, Lieutenant Commander Takahashi Kakuichi, the dive bomber commander, mistook the signal. The attack was supposed to begin when Fuchida fired a flare. But the commander of the fighter squadron had not seen the flare, so Fuchida fired another. Takahashi thought that meant the Americans were alert.[2]

The Americans were not only not alert, they didn't think a torpedo attack was even possible, because Pearl Harbor was too shallow to successfully launch aerial torpedoes. Admiral James O. Richardson, the previous commander of the Pacific Fleet, had rejected the suggestion that torpedo nets be placed in Pearl Harbor because "ships, at present, are not moored within torpedo range of the entrance."[3] Richardson was thinking of ship-launched torpedoes. He had not even considered the possibility that torpedoes might be launched from the air.

But Japanese planes were launching torpedoes, and the torpedoes were not, as expected, burying themselves in the mud under Pearl's shallow waters. One torpedo passed under Furlong's flagship, *Oglala*, and hit the cruiser *Helena*. The explosion burst *Oglala*'s seams. Two other torpedoes passed under the repair ship *Vestal* and hit the battleship *Arizona*, which was undergoing repairs. Three more hit the battleship *Oklahoma*. A third battlewagon, *West Virginia*, was hit several times. *Oklahoma* capsized. By controlled flooding, *West Virginia*'s crew managed to keep it upright as it settled into the mud of the harbor bottom.

The high-level horizontal bombers followed the torpedo planes. They were less accurate than the dive bombers, but their bombs, falling from a much greater altitude, had far more velocity and penetration. A bomb from one of the horizontal bombers penetrated *Arizona*'s deck armor and exploded in its magazine. The blast blew out the fires the other bombs and torpedoes had started and killed a thousand sailors. Half of all the navy men who died at Pearl Harbor—and almost half of all the people killed there—died in that explosion. The blast blew the men on *Vestal*'s deck into the water, including the repair ship's captain, Commander Cassin B. Young. The heat from the new fire on *Arizona* was so great that an officer on *Vestal* ordered, "Abandon ship." As sailors were starting to leave, an oil-covered figure just fished out of the harbor demanded, "Where the hell are you going?"

"We're abandoning ship."

"Get back aboard! You're not abandoning ship on me!" Commander Young roared.[4]

The Americans had been caught flat-footed, and Pearl Harbor was a disaster. There were some individual acts of heroism, though. *West Virginia*, hit by six torpedoes and several bombs, sank, but with its superstructure still above the surface, it continued to fight. Captain Mervyn Bennion, the skipper, was seriously wounded. Dorie Miller, a mess attendant, pulled the captain to shelter and manned an antiaircraft machine gun whose gunner had been killed. Miller was a mess attendant because he was African American and blacks at that time were limited to mess duty, so he had not received firearms training. Nevertheless, he managed to shoot down two of the attackers and kept firing until he was ordered to abandon ship.[5]

When Japanese Zeroes started strafing the naval air station at Kaneohe, some of the sailors shot back with infantry weapons. An aviation ordnanceman named Sands emptied a Browning Automatic Rifle (BAR) at a plane piloted by Lieutenant Iida Fusata.

Sands yelled, "Hand me another BAR! Hurry up! I think I hit that yellow bastard!"

Iida pulled out of his shallow dive and saw Sands standing in the field. Before the attack, he had told members of his squadron that if seriously hit, he would crash his plane into some enemy facility. He turned and dived at Sands. As he turned, the U.S. sailors on the ground saw a thin spray of gasoline coming from his plane.

With Iida's machine gun bullets kicking up dirt all around him, Sands emptied another BAR. The Japanese pilot returned to his squadron, pointed to himself and to the ground, then turned and again headed for the air station.

"Hey, Sands," a sailor yelled, "that sonofabitch is coming back."

Sands grabbed the nearest weapon, a bolt-action Springfield rifle, and began firing as Iida dived at him. On Sands's fifth and last shot, the Japanese pilot stopped shooting. His plane sailed over the field as if it were out of control and crashed into a low hill. Iida had been killed by a bullet from an obsolescent rifle wielded by a sailor who didn't care that what he was trying to do was supposed to be impossible.[6]

Exploits like those of Sands and Miller were merely bright spots in a very dark picture. The Japanese fliers had sunk or crippled every operational battleship in the U.S. Pacific Fleet. Five of the eight battleships at Pearl Harbor had been sunk, the other three damaged. (There was a ninth in the Pacific Fleet, the U.S.S. *Colorado*, that was undergoing repairs in Bremerton, Washington.) The Japanese had knocked out three cruisers, three destroyers, and four auxiliary craft. Most of the cruisers were, as is typical of cruisers, cruising. They were not at Pearl Harbor. The Japanese destroyed forty-six navy patrol bombers, twenty-one scout bombers, five dive bombers, thirteen fighters, and seven other navy planes. The army's Hawaiian Air Force lost sixteen bombers, two attack planes, fifty-six pursuit planes, and three observation planes. Most army and navy planes had been destroyed on the ground. Killed were 2,335 servicemen and 68 civilians.

Most of the Pacific Fleet would be out of action for months.

The Realization of Air Power

The commander of the attacking fleet, Nagumo Chuichi, an old battleship admiral, had limited faith in the air attack when he set out. He could not believe his success. From that point on, he was a confirmed air-power advocate, one of the few air-power enthusiasts among the world's admirals—even more enthusiastic than his superior, Admiral Yamamoto Isoroku, who had pushed the idea of attacking Pearl Harbor through a hostile naval bureaucracy.

An air attack on Pearl Harbor was Yamamoto's brainchild. He really had no business making plans: that was the duty of the Japanese Naval General Staff. Yamamoto was chief of the Combined Fleet. It was his duty to carry out the plans. But he saw that before Japan could carry out the army's plans to annex the Philippines, French Indochina, the Dutch East Indies, Burma, and possibly Australia, something would have to be done about the U.S. Pacific Fleet. No matter what his official duties were, Yamamoto, because of his intelligence and strength of character, was a formidable figure in the Japanese navy. At one point, he said that if he didn't get the ships he needed for the attack, he and his whole staff would resign. Other Japanese admirals had been reluctant to endorse the attack, because they thought it would be impossibly difficult.

Yamamoto knew it would be difficult. He turned to an officer very much his junior, Commander Genda Minoru, the Japanese navy's greatest torpedo expert. Genda recruited Fuchida Mitsuo, a brilliant aviator, to train torpedo plane pilots to release their "tin fish" while only a few feet above the water. Other officers in the Pearl Harbor team got Japanese torpedo makers to produce a new, shallow-running torpedo that would arm itself after a short travel. Horizontal bombers worked on their accuracy endlessly while Genda tried to find a bomb that could penetrate the heavy deck armor of American battleships. He ended up putting fins on sixteen-inch armor-piercing shells and using them as bombs.

American officials could not believe the disaster. "My God," said Navy Secretary Frank Knox. "This can't be true. This must mean the Philippines."[7] His army counterpart, Henry Stimson, was even more confused. "We know from interceptions and other evidence that Germany had pushed Japan into this," he said.[8] He knew no such thing. There were no such interceptions or evidence. Germany had no leverage over Japan. If anything, the leverage was in the hands of the

Japanese. Japan had joined the Tripartite Pact with Germany and Italy so it would have an excuse to snatch the Southeast Asian colonies of the defeated enemies of its "ally" Germany. France and the Netherlands were already out of the game, and it looked as if Britain might soon follow them. Nobody in Japan had any intention of giving Germany a share in this loot.[9]

American officials, however, were focused entirely on Europe. The U.S. Navy was attacking German U-boats well before Pearl Harbor. Before the attack, Stimson had been urging President Franklin D. Roosevelt to transfer *all* navy ships to the Atlantic. And Admiral Ernest J. King, as discussed in chapter 16, had grabbed all of the newest and fastest battleships, carriers, and cruisers for the Atlantic Fleet. That, it turned out, was a lucky break for the navy: its best ships were not caught in the Pearl Harbor surprise. Another lucky break was that the three aircraft carriers in the fleet were elsewhere. *Saratoga*, like *Colorado*, was at Bremerton, while *Lexington* was taking planes to Midway, and *Enterprise* was taking another batch of planes to Wake Island.

Big ships in the Atlantic, however, were practically useless. What was needed there were destroyers, destroyer escorts, and smaller vessels for convoy duty, as well as small aircraft carriers that could carry a few planes to locate and sink subs. But for a long time, King was reluctant to send his prized battlewagons and carriers to the Pacific, where they could do some good. Before Pearl Harbor, he fully supported an Anglo-American military conference that declared, "Since Germany is the predominant member of the Axis Powers, the Atlantic and European area is considered to be the decisive theater, and operations of United States forces in all other theaters will be conducted in such a manner as to facilitate that effort."[10]

But after an enemy attacked U.S. forces on U.S. territory, King began to have a change of heart, even if the enemy was not the "predominant" Axis Power. He still professed to believe in the "Europe first" policy, but he rejected the British demand that the United States remain on the defensive in the Pacific until Germany was beaten. As the Japanese began building new bases in captured territory in the South Pacific, King feared that they would push on into New Guinea,

Samoa, and Australia. The United States had to launch a counter-offensive. Even so, King only gradually released his hoard of capital ships in the Atlantic. General George C. Marshall and the U.S. Joint Chiefs of Staff strongly opposed a "diversion" in the Pacific. But long before the end of the war, almost all of the U.S. Navy's most powerful ships were engaged against Japan. The official line stayed "Europe first," but by the time Germany surrendered, the Japanese had been pushed back to their home islands, had no air-frame factories, had hardly any commercial shipping, had only remnants of their navy, and had no oil to fuel either ships or planes. And a large part of the Japanese army was stranded in places like New Guinea and Burma, imprisoned by the U.S. Navy as effectively as if they were in prisoner of war camps.

Through much of 1942, however, battleships in the Pacific were conspicuous by their absence. Of necessity, the United States abandoned the Jutland-style battle fleet formation and adopted a new one: the carrier task force. The early task force was built around one or two carriers protected by cruisers and destroyers. Later, battleships joined the task force and occasionally performed spectacularly. In late 1942, *South Dakota* shot down more than forty Japanese planes while it was protecting the carrier *Enterprise* off Guadalcanal. Occasionally, battleships were able to fight other battleships in the good old-fashioned way. The U.S.S. *Washington*, another new battleship in the Guadalcanal campaign, found the Japanese battleship *Kirishima* on its radar screen and opened fire. In seven minutes, *Kirishima* was out of action, and its crew scuttled it.

However, there were no battleships in the Pacific fleet in June 1942, when Yamamoto launched an expedition that he hoped would wipe out the U.S. Pacific Fleet once and for all.

Midway to Victory

The first carrier task force operations were hit-and-run raids. U.S. carrier planes hit Kwajalein, the Marshall Islands, Rabaul, and Japanese bases on the coast of Papua. On May 3–8, 1942, there was a major battle in the Coral Sea that involved a two-carrier U.S. task force and a Japanese fleet escorting troops to the southern coast of New Guinea. One American carrier, the U.S.S. *Lexington*, was sunk, and the other, the U.S.S. *Yorktown*, was damaged. *Yorktown* went back to Pearl Harbor and was hastily repaired. One Japanese carrier was so heavily damaged it would be out of action for months, and the second lost almost all of its planes. Tactically, because of the loss of *Lexington*, one of the oldest carriers in the U.S. Navy, the Battle of the Coral Sea was a Japanese victory. Strategically, it was an American victory. No Japanese warship ever again entered the Coral Sea. The Battle of the Coral Sea was notable for another thing. It was the first naval battle in which the opposing fleets never caught sight of each other.

Although it was a major battle, the Coral Sea did not have as much impact in either Japan or the United States as one carrier task force raid that inflicted only minor damage. Sixteen Army Air Force B-25s took off from the carrier *Hornet* and bombed Tokyo. Unable to land on the carrier deck because they were too big, most of them landed in China. The Japanese public was shocked. Japanese authorities believed the planes came from Midway Island, which prompted Yamamoto to devise a plan at least as ambitious as the Pearl Harbor attack. He would take Midway and, in the same operation, annihilate what was left of the U.S. Pacific Fleet.

His attack would open with a diversionary attack on the Aleutian Islands to draw the Pacific Fleet away. While the Americans were rushing north, Yamamoto's main force would occupy Midway, base planes there, and wait for the Americans to return. When that happened, Japanese submarines, Midway-based planes, carrier planes, and the main battle fleet would wipe them out. To the Aleutians, Yamamoto sent two light carriers, two heavy cruisers, and two troop transports. Meanwhile, the Japanese moved against Midway in waves. The first wave contained sixteen submarines. The second was Admiral Nagumo's task force with four big carriers—all the operational large carriers in the Japanese navy at that time. The third wave was most of the rest of the Japanese navy, commanded by Yamamoto himself. Yamamoto's flag ship was *Yamato*, one of the two largest battleships in the world. The other was its sister ship, *Musashi*. These two battlewagons made the famous German *Bismark* look like a cruiser. Each displaced 64,000 tons of water and carried nine 18.2-inch guns—the most powerful guns ever put on water. Their armor belts were sixteen inches thick.[11] Nagumo's carriers were to soften up the Americans and Yamamoto's battleships would finish them off.

Altogether, Yamamoto had 162 ships. The Pacific Fleet, under Admiral Chester W. Nimitz, had seventy-six. But a third of them were assigned to the North Pacific Force under the eccentric Admiral Robert "Fuzzy" Theobald. They never saw action—even in the Aleutians they were supposed to be guarding—because Theobald thought he knew the situation better than Nimitz's staff and had his ships in the wrong place. Nimitz had two task forces based in Pearl Harbor. The senior of his two subordinate rear admirals, Frank Jack Fletcher, had his flag on *Yorktown*. Because of the desperate situation of the Pacific Fleet, *Yorktown* was repaired in two days—a job that normally would have taken ninety days. The repairs were hasty, indeed. In normal times, *Yorktown* never would have been allowed out of the harbor. The other rear admiral, Raymond Spruance, had two carriers, *Enterprise* and *Hornet* in his task force.

Yamamoto had one problem. He couldn't imagine that the Americans would not fall for his Aleutian diversion. He didn't know that they knew his real destination. They knew because they had finally broken the Japanese naval code. Ever since Pearl Harbor, there has been an enormous amount of bilge written about the breaking of Japanese codes. The United States had not broken all of the Japanese codes before Pearl Harbor. American cryptanalysts had broken some of the diplomatic codes, but none of the military or naval codes. Before Midway, they had broken one of the naval codes. But, as we saw in the discussion of the Enigma codes (see chapter 16), breaking a code does not mean you can read all of an enemy's communications like a first-grade primer. Code keys get changed frequently, and even when the key is known, it takes a highly sophisticated computer to turn the message into something intelligible.

Before Midway, the Americans knew the main Japanese objective, but not details of the plan. And they did not know the exact locations of the Japanese ships. Fog hid Nagumo's task force from the eyes of

U.S. patrol plane crews. The Japanese, on the other hand, were sure the American ships were in Alaskan waters. They were also sure that *Yorktown* was on the bottom of the Coral Sea.

On June 4, 1942, Nagumo sent seventy-two bombers and thirty-six fighters to attack Midway. About ten minutes later, a U.S. Navy PBY patrol plane spotted a Japanese carrier. Then another plane saw the swarm of Japanese carrier planes heading for Midway and gave the alarm. Every plane on Midway took off before the Japanese arrived. The Marine Corps' old Buffalo fighters tried valiantly to shoot down the attackers, though they were outclassed by the Zeros. Regardless, the Buffaloes and the antiaircraft gunners did manage to shoot down a third of the attackers. Ten torpedo bombers—six navy and four army—from Midway then attacked the Japanese carriers. None scored a hit, and the Zeros shot most of them down. Next, sixteen B-17s, the army's wonder bombers, appeared. Before Pearl Harbor, General Marshall had assured President Roosevelt that no Japanese ships could attack Pearl Harbor because the B-17s would destroy them before they got close. This time, not one of their bombs hit a Japanese ship. During the whole war, no B-17 ever sank a Japanese warship. The U.S.S. *Nautilus*, a submarine, fired one torpedo at a Japanese carrier that missed. A storm of depth charges drove *Nautilus* away.

The attacks by the planes from Midway convinced Nagumo that the island needed more softening up. He had kept ninety-three planes in reserve so that in the unlikely event—he thought—that he saw any American ships, he could attack them. The planes were armed with torpedoes and armor-piercing bombs. He now ordered those planes to be rearmed with fragmentation and incendiary bombs.

Aboard *Enterprise*, Captain Miles Browning, Spruance's chief of staff, guessed that Nagumo would order a second strike at Midway. If planes from *Enterprise* and *Hornet* left now, they could catch the Japanese when they were refueling and rearming. Spruance considered the situation. The Japanese carriers were almost at the limit of his planes' range. It would be a risky operation. But the chance of catching the Japanese with their formidable Zeros on deck was worth the risk. He ordered the attack. Fletcher delayed launching *Yorktown*'s planes in case more Japanese were located, but he sent his bombers and fighters off two hours later.

Meanwhile, a Japanese scout plane had sighted the American ships. Nagumo changed course to meet the new enemy and had his reserve planes rearm to deal with the ships. The Americans did not know of this change of course. The U.S. torpedo bombers, dive bombers, and fighters, traveling in separate groups, were headed in the wrong direction. Scouting the area, *Hornet*'s torpedo squadron located the Japanese carriers. There were four, surrounded by two battleships, three cruisers, and eleven destroyers. Not all the Japanese Zeros were being refueled. Some Zeros dived on the wave-skimming torpedo planes and, with the gunners on the ships, shot all the American planes down. Only one man survived from Torpedo Eight. Next, the *Enterprise* torpedo squadron appeared. It lost ten of its fourteen planes and didn't score a hit. *Yorktown*'s torpedo squadron, which had learned the location of the Japanese task force and didn't lose time scouting the sea, appeared next. It, too, lost ten planes without scoring a hit.

Nagumo's task force, it seemed, had practically won the Pacific war. It had beaten off attacks by land-based bombers and a submarine and had shot down almost all of the Pacific Fleet's torpedo planes.

This Japanese triumph lasted exactly 100 seconds.

At 10:26 A.M., Lieutenant Commander Clarence McClusky, leading two dive bomber squadrons from *Enterprise*, saw the carrier *Kaga*. Then he saw a second, *Akagi*. He had one squadron follow him when he dived on *Kaga*, and sent the other squadron, under Lieutenant W. E. Gallaher, to attack *Akagi*. The Zeros, all at low altitude where they had attacked the torpedo planes, could do nothing to stop the dive bombers. *Kaga* took four hits that killed everyone on its bridge and set it ablaze from stem to stern. An internal explosion sent it to the bottom. One bomb exploded in *Akagi*'s hangar, detonating its torpedoes. Another wiped out all the planes trying to refuel on its deck. *Akagi* was Nagumo's flagship. He transferred his flag to the cruiser *Nagara* after ordering the crew to abandon ship. *Yorktown*'s dive bombers, under Lieutenant Commander Maxwell Leslie, arrived in time to see a third carrier, *Soryu*, preparing to launch its Zeros. Leslie's planes scored three direct hits, wiping out the Zeros and turning *Soryu* into an inferno. Then the sub *Nautilus* returned and shot three torpedoes into *Soryu*. The big carrier broke in half and the two pieces went to the bottom in a hissing cloud of steam.

Nagumo didn't give up; he sent planes from his remaining carrier, *Hiryu*, to attack *Yorktown*. "Waltzing Matilda," as she was known in the navy, was far from tip-top condition. Three bomb and two torpedo hits were too much for its jury-rigged repairs. It was abandoned except for a repair crew that was trying to save it. The destroyer *Hammann* sailed beside it to furnish electric power and the minesweeper *Vireo* towed it back toward Pearl Harbor. It looked as if *Yorktown* would live to fight another day. Then a Japanese submarine torpedoed the carrier, the destroyer, and the minesweeper, and all sank. Before that, dive bombers from *Yorktown* had located *Soryu*. They flew to *Enterprise* to refuel and, with Gallaher's squadron from *Enterprise*, returned and sank *Soryu*. All of Japan's operational heavy carriers had been sunk.

Yamamoto said he would attack with his battleships. Nagumo argued against that, and Yamamoto removed him from command. Then he had second thoughts. The Americans now controlled the air. To continue toward Midway might well mean the end of all Japanese naval power. He ordered his ships to return to Japan.

The Japanese navy came to realize that there was a new capital ship. The battleship was no longer queen of the seas. The new queen was the aircraft carrier. A third super battleship, planned to be a sister to *Yamato* and *Musashi*, was converted to a carrier. *Sinano*, at 59,000 tons displacement, was the largest carrier in the world when it made its maiden voyage in November 1944. But it never finished the voyage. On November 29, the U.S. submarine *Archerfish* sank *Sinano* before it could send a plane into combat. A month before that, the mighty *Musashi* took nineteen aerial torpedo and seventeen bomb hits—enough hits to destroy most navies—and went down. Its sister ship, *Yamato*, lasted until April 7, 1945, when, after two hours of unrelenting aerial bombardment, the world's most powerful battleship went to the bottom.

19

War Against the Home Front

Nuclear bomb explodes over Nagasaki, Japan, August 8, 1945. In World War II, civilian populations became military objectives. (Office of War Information photo, National Archives.)

Apostles of Air Power

IN 1921, Giulio Douhet, an obscure Italian general and artilleryman-turned-airman, published the book *The Command of the Air*. The most important thing in the next war, he contended, would be to be "in a position to prevent the enemy from flying while retaining the ability to fly oneself."[1] It would then be possible, he said, to bomb the enemy into submission with little or no help from the earthbound fighting forces. By bombing factories, distribution centers, roads, and railroads, an enemy country would be paralyzed, and the army and navy would have nothing to do but some mopping up. In a later edition of his book, in 1926, Douhet argued for an independent air force—as independent of the ground forces as the navy is from the army. Airplanes, he claimed, are weapons of unlimited offensive power and no defense against them is possible.[2] He strongly advocated bombing cities to break civilian morale. That, in the long run, he argued, would save more lives.

Douhet's book was not published outside of Italy until after his death in 1930. Meanwhile, two other generals, Hugh Trenchard in Britain and William Mitchell in the United States had begun preaching the supremacy of air power and the need for an independent air force. According to John Keegan and Andrew Wheatcroft, "Trenchard's life demonstrates how on rare occasions a revolution—in this case a technical one—may elevate a man of humdrum career to the heights of power almost overnight. At the age of thirty-nine, Trenchard was a major without any prospects."[3] He had failed to get a commission in the regular army and navy, and he took the militia examination, his last hope, three times before he passed. After a number of years in the army, he learned to fly and, because of his age, became the most senior officer in the new Royal Flying Corps. From that position, he began lobbying for an independent air force. He became the first marshal of the Royal Air Force (RAF) and established its orientation toward strategic bombing. He agreed with Douhet's ideas long before he saw his book. If there was to be an independent air force, it must have an independent role to play. Strategic bombing was something an air force could do but an army or navy could not, and it did not involve cooperation with either of the surface forces.

The third apostle of air power was the American, Billy Mitchell. Mitchell has become a kind of hero in the United States, mostly for people who have never understood what he was preaching or what happened to him. Mitchell had been in charge of American air operations in World War I and was promoted to brigadier general in 1921. On the western front, his fliers were extremely successful in supporting the ground troops by strafing and bombing. They also interdicted enemy supplies and reinforcements. After the war, the Army Air Corps became responsible for coast defense. Mitchell demonstrated air power for coastal defense by sinking the old hulk of a former German battleship and later two decommissioned American battleships. For this, he has been hailed as a prophet whose warning was ignored until Pearl Harbor, twenty years later.[4] This view ignores three facts: (1) the ships sunk were stationary, decrepit, and had no antiaircraft guns; (2) in World War II, no high-level bomber, the kind Mitchell used, ever sank a moving battleship; and (3) on occasion, as with *South Dakota* in the Guadalcanal campaign, battleships protecting aircraft carriers shot down scores of much better planes than Mitchell's without being sunk. For Mitchell's adherents to say his demonstration predicted what would happen at Pearl Harbor is like saying that the first gunpowder weapons predicted what machine guns would do in World War I.

Mitchell strongly supported what today we call tactical air support of ground forces, but that concept needed little assistance. All military authorities were in favor of air support. But like Douhet and Trenchard, he also argued for strategic bombing. Like them, he believed that a strong, independent air force was about all the military force a country needed. As time went on, he became more and more strident in his arguments, finally accusing the U.S. military chiefs of being incompetent, which is a bit strong for a serving officer. He was court-martialed and, contrary to popular belief, was not discharged. He was suspended in rank for five years. He quit the army in disgust.

Meanwhile, the doctrine of strategic bombing gained new converts, especially in Britain. Both J. F. C. Fuller and Basil H. Liddell Hart, the armor advocates, adopted it.

"The result of warfare by air will be to bring about quick decisions," Fuller wrote. "Superior air power will cause such havoc in the opposing country that a long drawn-out campaign will be impossible."[5]

Havoc among the civilian population was something Fuller was all in favor of, believing, like Douhet, that it would save more lives in the long run. He hailed the use of gas as an aid to that end. "It is unnecessary that . . . cities be destroyed, in the sense that every house be leveled with the ground. It will be sufficient to have the civilian population driven out so that they cannot carry on their usual vocations. A few gas bombs will do that."[6]

Fuller became downright enthusiastic about gas. He said gas could be used to obliterate all life on the frontiers while bombers destroyed the enemy's industry, transportation, and government functions.[7]

Liddell Hart, whose military career was destroyed by mustard gas, was much less enthusiastic about chemical warfare, but he, too, advocated the destruction of an enemy's cities to eliminate his will to continue fighting. Writing in 1925, Liddell Hart predicted the effects of bombing on cities, including "the slum districts maddened by the impulse to break loose and maraud."[8] His was a fairly common misconception among the English upper classes, fostered by what had happened in Russia in 1918—that all the European lower classes were a human bomb that needed only some disaster to make them "break loose and maraud." In 1941, Winston Churchill was looking for a German revolution when he wrote, "There is one thing that will bring [Adolf Hitler] down, and that is an absolutely devastating exterminating attack on the Nazi homeland." On February 14, 1942, the RAF staff directed that operations "should now be focused on the morale of the enemy civilian population and in particular of industrial workers," an order that Air Chief Marshal Sir Charles Portal explained further, "the new aiming points are to be built-up [residential] areas, not, for instance, the dockyards or aircraft factories."[9]

A later British convert to bombing cities was Air Marshal Arthur "Bomber" Harris. In World War II, Harris decided that bombing key targets like railroad marshaling yards and certain industries was ineffective. He instituted "carpet bombing," that is, using as many planes as possible to lay bombs over cities. He reasoned this type of bombing would probably hit targets of military value and would depress civilian morale.

When war began, the theories of Douhet et al. would finally be tested.

The Battle of Britain: Opposing Forces

Surprisingly, Germany, which had demonstrated its utter ruthlessness in so many ways, including the bombing of the undefended cities of Warsaw and Rotterdam, had not subscribed to the benefits of city bombing. Hermann Göring, the leader of the Luftwaffe, had been an ace fighter pilot in Manfred von Richtofen's Circus. He saw fighter pilots as "knights of the air." Fighter pilots in World War I (and later in World War II) often engaged in ground support, so Göring saw Stuka pilots, too, as brothers. But to him, other bomber pilots were "truck drivers." Strategic bombing was not one of the Luftwaffe's top priorities.

Things changed after the fall of France. Hitler expected Britain to ask for terms, after which it would become Germany's junior partner in his dreamed-of crusade against the Soviet Union. He had some basis for his belief. Negotiations with Germany were discussed in the British war cabinet on May 27, 1940, before the evacuation of Dunkirk. Lord Halifax, the foreign secretary, recommended giving Benito Mussolini Malta and Cyprus and a share in the running of Egypt so he would become an intermediary between the British government and Hitler. Neville Chamberlain, the former prime minister, supported Halifax. Even Prime Minister Churchill said that if Germany would settle for the return of its former colonies and the overlordship of central Europe, peace might be possible.[10] The next day, however, Churchill changed his mind.

In spite of rejection, Hitler continued to hope. Finally, on July 16, he ordered his military chiefs to prepare plans for an invasion of

Britain. And he said he wanted his forces to be ready to move by mid-August. Hitler had seen plenty of land warfare in World War I, although from a worm's-eye view. His knowledge of sea warfare was nil. He had a most impressive army, but his navy was no match for Britain's—the second largest in the world. Obviously, the success of the projected invasion would depend on his air force.

"The Fuhrer has ordered me to crush Britain with my Luftwaffe," Göring told his generals on August 1, about three weeks after the Luftwaffe had begun bombing Britain. "By means of hard blows, I plan to have the enemy, who has already suffered a crushing defeat, down on his knees in the nearest future, so that an occupation of the island by our troops can proceed without any risk."[11]

The trouble was that his Luftwaffe was not prepared for that kind of war, which was principally Göring's fault. The Luftwaffe had been designed primarily to give the army tactical support, not crush an enemy from the air, Douhet-style. But Göring had never been interested in the problems of bombers. He looked at a map and saw that planes would be able to hit Britain from Norway, Denmark, the Netherlands, Belgium, and France. These attacks should overwhelm the British, who had lost 430 top-of-the-line fighter planes in France, along with most of their army equipment. Göring told the generals that bringing Britain to its knees would take no more than four days.

Theoretically, German planes could attack Britain from the whole Atlantic coast of Europe north of the Iberian Peninsula. Actually, the premier German fighter, the Bf 109, had a range of only 125 miles. Operating from the bases closest to England, it could spend no more than twenty-five minutes in enemy territory. Germany had bombers with more range, of course, but without a fighter escort any bomber crews flying from Norway to England would never return alive. This was because of another factor Göring had ignored: the British fighter production. In 1940, Messerschmitt was producing 140 Bf 109s and 90 Bf 110s a month. At the same time, Vickers and Hawker were turning out 500 Spitfires and Hurricanes a month. The Spitfires and the Bf 109s were the best fighters in the world. The Hurricane was a little slower and less maneuverable than the Spitfire, but it was far better than the Bf 110.[12]

Göring's biggest miscalculation was that he didn't know that the British had radar stations covering all Britain's eastern and southern coasts. Robert Watson-Watt had invented radar, and the British had a far more sophisticated system than anything the Germans had. Radar could pinpoint attacking aircraft, and the British Fighter Command could hit them with planes from all over the island.

The Battle of Britain: Birth of the Blitz

In spite of Göring's boastful words, the Battle of Britain began not with a bang but a whisper. It opened on July 10 with some raids on England's southern coast that sank some merchant ships. There were no hits on the Royal Navy, which should have been the prime objective of a ground-support force like the Luftwaffe. The Germans sent between twenty and thirty planes on each raid. By July 31, they had lost 180 planes; the British, 70. The Stuka, which had terrorized troops in Poland and France, was withdrawn after a couple of weeks in combat. Against enemy planes, it was a flying coffin. On August 13, the Germans shifted their targets from ships to air bases and aircraft factories. Somehow, they managed to hit only one radar station, but it looked as if they might wear down the RAF by impeding fighter production and destroying fighters on the ground. They sent their planes over in waves, so that while the British fighters were refueling, the second wave would catch them on the airfield.

Then some German pilots had a panic attack and created the first Douhet-style war. On the night of August 24–25, ten German bombers, which were supposed to bomb a fuel storage area, jettisoned their bombs and fled back to the continent. The bombs landed in the heart of London. The British were outraged and demanded that Berlin be bombed. The RAF had been created to carry out strategic bombing, but it was even less prepared to do it than the Luftwaffe. It didn't have the planes, and during 1940 the British factories had been concentrating on producing fighters, not bombers. The Berlin raid was totally ineffective. But now the Germans were outraged and retaliated. They switched from military targets to London and other British cities.

This was what the British called "the blitz"—something quite different from the blitzkrieg. It wasn't "lightning war" unless you consider the deliberate murder of civilians war. And there was no lightning in it. It went on and on, and the British kept fighting. Contrary to the predictions of Douhet, Trenchard, Mitchell, Fuller, and Liddell Hart, the air war did not create panic, confusion, and crush the will of the people. All through the war, 60,000 people, almost all of them civilians, were killed in the blitz.[13] And it affected the outcome of the war not one bit. So far, the theories of Douhet and his disciples had been proved utterly false. Now, the British, and then the British and the Americans, were to give them a sterner test.[14]

Retribution

The British air war against Germany was handicapped by the fact that Germany was separated from Britain not only by the English Channel, but also by most of western Europe. British bombers would have to fly over France, Belgium, or Holland, all of which were loaded with German air bases and the excellent Bf 109 fighters. They'd have to run a gauntlet of Messerschmitts. Even though the British had no long-range fighters that could protect their bombers from the Messerschmitts, they still managed to bomb Germany. Some in England wondered if it was worth the price. During 1941, more British aircrewmen were killed in the raids than German civilians, and producing bombers to keep up the campaign diverted resources from other war production.[15] The strategic bombing campaign also took planes away from antisubmarine patrol, something far more important militarily.[16]

Because they had no fighter protection, the British bombers took to night raids. Night in the sky, especially over a blacked-out continent, is very black indeed. Enemy fighters had trouble finding the bombers, but the bombers also had trouble finding their targets. They often missed entire cities. The British eventually developed navigational aids, using radio beams and radar to help find targets. But the most effective method was using "pathfinders"—the fast, high-flying Mosquito fighter-bombers that flew out of the range of antiaircraft guns and dropped flares over the target city. The German interceptors also improved their night navigation and seldom missed finding

Planes, Firepower and Fire Effect

Could a single bomber really have the firepower of two or three fighters? Why not? When the B-17 had five .30 caliber machine guns, all American fighters had two .30 machine guns. When the B-17 had ten .50 caliber machine guns, the top enemy fighter, the FW 190, had two 20 mm and two 7.92 mm machine guns. Four into ten is two and a half. We are talking here about firepower, not fire effect. The FW 190, with a speed of around 400 miles per hour, greatly outclassed the B-17, which could barely make 300 miles per hour. The FW could bring all its guns to bear on a target while the Flying Fortress could use only six or eight of its ten guns, depending on the direction of the target. On the other hand, the B-17 could absorb far more punishment than any fighter.

According to some caclulations, the .50 caliber Browning aircraft machine gun could, because of its rate of fire, put a greater weight of projectiles on a target in a given time than either a 20 mm or a 37 mm automatic cannon. So, the USAAF thought the B-17s could take care of themselves without fighter escorts. They didn't include range in their calculations. The auto cannons outranged the .50s, and the German rockets outranged them more.

Nevertheless, a German fighter pilot attacking a Flying Fortress would be well advised to have paid up life insurance. The kill rate wasn't that heavily in favor of the Germans. Each B-17, though, was far more expensive in dollars and lives than any fighter.

bombers as the war went on. The somewhat clumsy Bf 110 became a highly successful night fighter because it could carry heavy armament and didn't have to dogfight with British fighters.

British raids increased in intensity, however. Harris got bigger and faster planes that could carry more bombs. He began using incendiaries on a large scale. At Hamburg, he created his masterpiece. For four nights in a row, RAF bombers hit the big German port city. They created a "firestorm," perhaps the first created in war since the burning of Magdeburg in 1631 during the Thirty Years' War. The fires in the center of the city became so hot they caused a tremendous updraft, which resulted in cyclone-like winds on the surface and sucked all the oxygen out of the air that surrounded the inferno. Some people suffocated in their bomb shelters. Altogether, 30,000 civilians died in those four days in that one city.

In 1943, the U.S. Eighth Air Force joined the strategic bombing campaign. In the 1920s, Douhet contended that there was no defense against bombers. The German raids on Britain showed that bombers without fighter protection are in deep trouble. The British raids on Germany proved that conclusively. But the Americans thought they had the answer. It was called the B-17, a fast, well-armed heavy bomber. When the plane was introduced in 1935, it could fly at least as fast as any American fighter, and with the firepower of two American fighters. By 1942, when it entered the war, it had ten .50 caliber machine guns, which gave it the theoretical firepower of two Focke-Wulf 190s or Messerschmitt bf 109s. At the time, the USAAF considered the .50 machine gun the equal of the 20 mm cannon because of its faster rate of fire. The B-17 also had the Norden bombsight, perhaps the most closely guarded U.S. military secret.

The American strategy was to use this formidable machine on daylight raids against "bottleneck" industries. Ball bearing production was a bottleneck industry; everything needed ball bearings, but there were only a few places that produced them. American economic analysts said wiping out German ball bearing production would cripple the German war industry. The first target of the B-17 Flying Fortresses was the ball bearing works at Schweinfurt. Actually, the Germans also had a ball bearing plant at Regensburg, and they imported large quantities of the little steel balls from Sweden.

The Schweinfurt raid was a disaster. Fighter planes had become much faster since the Flying Fortresses had been introduced. There was no way a B-17 could outrun a Bf 109. And all those .50 caliber machine guns were outranged by the automatic cannons that flew on the German fighters. Later, the fighters added rockets to their cannons and fired them at the big bombers as if they were U-boats attacking a fleet. Of the 229 Flying Fortresses that took part in the raid, thirty-six were shot down. A similar raid on another ball bearing factory in Regensburg resulted in the loss of twenty-four more Flying Fortresses. Furthermore, another 100 B-17s were damaged in these unsuccessful attempts to deprive Germany of ball bearings. The Eighth Air Force suspended deep raids for five weeks. The Flying Fortress, the plane expected to fulfill the Douhet ideal of a bomber that could fight off fighters, did not meet expectations. It could mix it up with fighters, but losses in such encounters were not acceptable.

Salvation came in the shape of long-range fighters. By fitting drop tanks to the twin-engine P-38 and the huge P-47, these fighters could escort bombers almost anywhere in Europe but eastern Germany. Then the Americans introduced the P-51—a fighter that could shepherd the bombers anywhere and outperform the German fighters on their home ground. The all-African-American 332nd Pursuit Group, flying P-47s and later P-51s, set a record while escorting U.S. bombers—not a single one of the bombers they guarded was ever shot down by a German fighter. The Americans returned to their daylight raids on key industries, concentrating on airplane factories and synthetic oil plants. Because of these raids, German oil production between March and September 1944 declined from 316,000 tons to 17,000 tons.[17]

During these raids, the U.S. bombers relied more on massive numbers of planes than on the vaunted Norden bombsight. The Norden bombsight, which was supposed to let a bombardier drop a bomb into a pickle barrel from 10,000 feet and keep the U.S. shores safe from enemy ships, began to look like the most overrated military secret since the Montigny mitrailleuse. The U.S. Army Air Force also joined the RAF on night raids, where the bombsight was useless. Even though these raids disrupted transportation and wrecked factories, they mostly just destroyed houses and killed people. In all the raids on the industrial Ruhr area, 87,000 people were killed; in Berlin, 50,000; in Cologne, 20,000; and in Magdeburg, 15,000. Altogether, 600,000 German civilians were killed in air raids and another 800,000 seriously injured.[18]

Although German synthetic oil production was seriously affected, weapons production did not seem to be until Germany's enemies were already in *der Vaterland*. In 1942, the British dropped 48,000 tons of bombs, and the Germans produced 36,804 heavy weapons (artillery, tanks, and planes). In 1943, the Americans and British dropped 207,600 tons of bombs, and the Germans produced 71,693 heavy weapons. In 1944, the Americans and British dropped 915,000 tons of bombs, and the Germans produced 105,258 heavy weapons.[19] The value of all that bombing remains questionable. It certainly did not break the German morale, any more than the earlier German campaign broke British morale. Nor would the even more horrible campaign against Japan break the Japanese morale.

The Big One

While the ground war was raging in France, the Low Countries, and Russia, and while ever-larger air raids were pounding Germany, Japan was being pushed out of the Marianas, Saipan, and the Philippines; the once-powerful Japanese navy—practically tied with Britain's as the second-largest in the world—had been virtually wiped out. The United States had established a base on Saipan and reestablished bases on Guam and in the Philippines. From these bases, it sent a new plane, the B-29 Super Fortress, to blast the Japanese home islands. The bombers annihilated Japan's aircraft industry and most of its oil reserves, and they did not neglect the cities. For these, the Super Fortresses' main weapons were incendiary bombs—the old-fashioned thermite-and-magnesium bombs and a new type containing napalm (jellied gasoline). Japan, a land of mostly wooden houses, was extremely vulnerable to incendiary attack. Just before midnight on March 9, 1945, B-29s hit Tokyo and Yokohama and created the greatest firestorm in history. The Tokyo fire department reported that the raid had killed 97,000 people, injured 125,000, and left 1.2 million homeless.[20]

The bombings went on. Nagoya, Kobe, and Osaka were destroyed in firestorms. The updraft from these blazes tossed the huge bombers so violently that their crews' helmets were ripped off. But Japan fought

on. The Americans took Okinawa, which for centuries had been practically one of the Japanese home islands. It was a costly victory. The
army and marines lost 7,600 killed in the ground fighting; the navy,
5,000, from kamikaze attacks. Japanese dead came to 110,000. Only
11,000 surrendered.[21]

Japan got a new prime minister, Admiral Suzuki Kantaro, who
wanted peace. Suzuki tried to enlist Josef Stalin as an intermediary
with the Americans and British, but Stalin saw an opportunity to grab
some spoils on the cheap—just as the Japanese had when they signed
the Tripartite Pact. At Potsdam, on July 26, 1945, Stalin, Churchill, and
Roosevelt declared that Japan must surrender unconditionally or face
complete destruction. The Japanese people feared that their divine
emperor would be executed. Undersecretary of State Joseph C. Grew
urged that the Japanese be told that the emperor would not have to
abdicate. Secretary of War Henry Stimson agreed, while Secretary of
State James F. Byrnes opposed such notification. It was never sent. It
didn't really matter, because the Japanese military preferred death to
surrender and thought the rest of the country should as well. Suzuki
said his government would ignore the Potsdam terms.

The decision to use nuclear weapons belonged to President Harry
S Truman alone. Dwight D. Eisenhower, told during the Potsdam Conference that the atomic bomb would be used, said he believed that it
was "completely unnecessary."[22] Douglas MacArthur later said he
agreed with Eisenhower, but that he hadn't been consulted.

On August 6, 1945, a B-29 dropped the world's first nuclear bomb
to be used in war over Hiroshima and killed 78,000 people. Hiroshima
was a middle-sized city of no particular importance. All the bigger
cities had been virtually destroyed. On August 9, another nuclear
bomb exploded over Nagasaki, killing another 25,000. Japan's air
power had diminished so much the bombers attacked in daylight and
needed no fighter escort. Japan lay open to attack. The same day as
the Nagasaki attack, the Soviet Union sent three army groups into
Manchuria. Stalin knew he had to move fast if he wanted to make a
profit. He was right. Japan sued for peace on August 10.

Nuclear weapons have not been used since that day in 1945. The shadow of the nuclear bomb, however, has influenced all international relations ever since. The fact that it *could* be used has affected all military decisions since the end of World War II.

The strategic bombing in World War II, even apart from the atomic bombs, was unique in modern warfare. Even in the awful European wars of religion, such events as the sack of Antwerp and the burning of Magdeburg were regarded at the time as atrocities. Never since the days of Tamerlane and Genghis Khan had innocent civilians been wiped out with such casual disregard. It wasn't the kind of breathtaking evil as that of the Nazis' slaughter of 12 million Jews, Slavs, Gypsies, and others they considered undesirable—people killed just because they existed. Even the most callous must have known this was wrong. But the men who flew the bombers—and at great risk to their own lives—had no sense of wrongdoing. Their people hailed them as heroes. Killing enemy men, women, and children had become a military objective.

As with the excesses of the wars of religion, the World War II air campaigns produced a reaction. Now, we have weapons capable of guiding themselves to strictly military objectives. We call them "smart weapons," and they are featured in one of the latest turning points of warfare.

20

High Tech

Flying under radar control with a B-66 Destroyer, F-105 jets bomb a target hidden by low clouds in the southern panhandle of North Vietnam, June 14, 1966. (U.S. Information Agency photo, National Archives.)

Bolts from the Blue

EARLY IN THE MORNING of January 17, 1991, weary Iraqi air defense technicians peered at their radar screens, as they had been doing for weeks. They knew that an American-led combination of powers had been massing troops on the border of Saudi Arabia and Iraq's "lost province" of Kuwait. There was nothing on the radar screens.

247

The technicians were alert, even wired, this morning, though. They knew something was probably going to happen. At around 2:30 A.M., American helicopters had knocked out two radar sites near the Saudi border. Ten minutes later, the Nukhayb Intercept Operations Center went dead. But there was nothing on the screens.

At exactly 3:00 A.M., the Baghdad communications center blew up. Simultaneously, one-ton bombs struck command posts all through Iraq. The Gulf War had begun with a succession of strikes by an unseen, unheard enemy.

The 2,000-pound bombs that struck the communications center and the command posts were laser-guided weapons released by F-117 stealth fighters flying too high to be heard by listening devices and invisible to Iraqi radar. Those bombs were followed by Tomahawk cruise missiles fired from submarines and surface ships in the Persian Gulf and the Red Sea. The Tomahawks flew low, guided by a program installed in them and continuously checked by a radar on the missile that referred to landmarks on the flight path. The Tomahawks hit Baghdad's electrical transformer stations and exploded, throwing out carbon filament wires that shorted out the transformers. Power in most of Iraq failed. In the air defense stations, emergency generators took over, so the radar screens stayed active. But they showed nothing.

Then they saw it: two air armadas flying toward Baghdad. One came from Saudi Arabia, the other from American aircraft carriers in the Red Sea. Each air fleet contained electronic jammer planes and fighter-bombers. The jammers would interfere with the ground radar beams, forcing the Iraqis to turn up the power in order to see anything. That would make each radar transmitter a prime target for the weapons carried by the fighter-bombers—High-Speed Anti-Radiation Missiles (HARMs). Just to confuse the Iraqi defense, the planes also dropped twenty-five tactical air-launched decoys to give the radar operators something to look at. Then the HARMs hit the radar transmitters and there was nothing more to see.

While this was going on, another wave of F-117s attacked the air defense command posts. They released sixteen laser-guided bombs, ten of which hit within one meter of the center of their targets. For the rest of the Gulf War, Iraq's air defense would remain blind.[1]

And all this happened more than a decade ago.

The electronic revolution began affecting the military in a big way about the same time it began changing civilian life. Electronics, combined with nuclear power (to drive ships, not destroy cities), has radically changed warfare. Let's take a quick look at some—by no means all—of these changes, changes in the way war is fought at sea, on land, and in the air.

War at Sea

Nuclear power is the most obvious change in modern navies. It has been applied mostly to aircraft carriers and submarines. Aircraft carriers of the *Nimitz* class displace 95,360 tons. This simply dwarfs World War II–era ships. It's one and a half times the size of Japan's *Musashi*-class battleships, the ships that had America's naval planners on edge since the 1930s. It's almost three times the size of our largest battleships in World War II. There is simply no way a Nimitz-class carrier could squeeze through the Panama Canal. A huge carrier like *Nimitz* is horribly vulnerable, but given the conditions of the world today, it's absolutely necessary. Nimitz is vulnerable because it's such a big basket to put our eggs in, and there are a number of potential enemies of big aircraft carriers—enemies that are also powered by nuclear engines. But it's necessary because a huge carrier is the fastest way to meet U.S. military commitments anywhere on the globe. Carriers can carry a variety of aircraft—many types of planes, helicopters, and vertical-takeoff-and-landing planes like the British Harrier.

The most potent enemy of the huge carrier is the nuclear-powered submarine. World War II confirmed that the aircraft carrier was the new capital ship. In the postwar period, the submarine is challenging the carrier for this title. These new submarines are vastly bigger and stronger than those of either World War I or II. They even look different. They are completely streamlined, like a fish or a whale, because they're designed to run underwater almost all the time. They are faster underwater than most vessels are on the surface, and they can dive to depths submariners of the two world wars would have thought impossible. They can, as in the Gulf War and the later Iraq War, launch cruise missiles. Other subs can launch intercontinental ballistic missiles like the Polaris, and they can do it while submerged. Since about three-quarters of the earth's surface is water, there is no place in the world that is safe from attack by a submarine-launched missile.

Some nuclear subs are designed to hunt down and destroy other subs as well as surface ships. They carry a variety of weapons to do that. There are torpedoes, of course, but they're far more sophisticated than most of the "tin fish" of World War II. Some torpedoes trail a long, thin wire so the submarine crew can guide them to their targets. Others home in on the noise of the target vessel. Others are programmed to search for targets and to return and try again in case they miss the target. One submarine-launched antisub weapon is a rocket that breaks the surface of the water and flies to where detectors have located an enemy submarine. There, it drops a depth charge on the enemy submarine.

Most modern naval surface ships are basically armed with rockets, antiaircraft, or surface-to-surface missiles. Big guns are a thing of the past. Cruisers were once loaded down with as many as fifteen six-inch guns or eight to twelve eight-inch guns (the *Alaska* class carried twelve-inch guns), in addition to many antiaircraft guns. In the Iraq War, they carried two automatic five-inch guns. In most cases, the guns fired from unmanned, remotely controlled turrets, like the U.S. Navy's Mark 45 system.[2]

War on Land

The artillery used in the Iraq War would not seem strange to anyone who had been around the big guns in World War II, Korea, or Vietnam. But some of the fire-control methods might seem little short of miraculous. For instance, if an Iraqi gun fired on Americans, U.S. counterbattery radar could pick up the shell in flight, calculate its trajectory, and pinpoint the location of the gun that fired it. "In some cases," say Williamson Murray and Robert H. Scales Jr., "the Coalition had rounds on the way to an Iraqi unit before that unit's first round had landed."[3] Most modern artillery is self-propelled.

Rockets continued to play a major role in artillery. The biggest Coalition field artillery rocket launcher was the M 270 MLRS, a tracked vehicle that could fire twelve 227 mm rockets in less than a minute and hit targets up to twenty-eight miles away. Some of the rockets from the MLRS were filled with some 600 bomblets. When the rocket exploded, these were spread out over an area the size of a football field. The rocket launcher was an extremely effective antipersonnel weapon.

The biggest advance in artillery rockets, though, came in antimissile weapons. In the Gulf War, the Patriot antimissile missile failed to live up to its press notices. It did not, as earlier reported, stop many Iraqi Scud missiles. The earlier Patriot had been designed to explode when it neared an enemy missile and destroy it with the metal fragments blasted out by the explosion. The trouble, say Murray and Scales, was that the Scud near the end of its trajectory was traveling so fast it could outrun those fragments.[4] The new Patriot rammed directly into the incoming Scud and had a high success rate.

Modern tanks are bigger, heavier, more powerful, and more heavily gunned than those of earlier wars. One of the biggest changes is that many of them, like the M1 A1 Abrams, have a smoothbore main gun. The gun, of 120 mm in the Abrams's case, does not spin the shell so the jet from the explosion of a shaped-charge shell will not be diminished. Projectiles from the Abrams's gun are stabilized by fins. In addition to shaped-charge (HEAT, or high-explosive antitank) shells, tank and antitank guns make increasing use of sabot shot. The "sabot" is a light casing around a solid shot of much smaller diameter than the bore of the gun. As it is driven by the propelling charge for a much heavier shot, it leaves the muzzle at a tremendous velocity. The sabot immediately falls off, and the shot continues on to the target. U.S. sabot shot are generally made from depleted uranium (DU), an extremely hard and heavy metal. Most of the radiation is gone from DU, but when it strikes something hard, it gives off sparks of extremely high temperature. So it can both penetrate armor and cause fires or explosions.

The frontal armor of modern heavy tanks is quite thick, and on

American and British machines it is of the Chobham type, which has a nonmetallic component to foil the acetylene torch effect of a shaped-charge explosion. The great advantage of U.S. and British tanks during the Iraq War was not their guns or armor, however, it was their sensors. There were a variety of these devices that let Coalition tanks locate and target enemy tanks before the enemy knew there were hostile tanks in the area. The sensors gave the Americans and British an advantage at any time, but they provided a tremendous edge at night or in sandstorms or fog.

Modern infantry seldom march great distances. They ride, and when they're near the firing line, they ride in armored personnel carriers (APC). The U.S. Army's current APC is the Bradley Fighting Vehicle. The Bradley, armed with a turret-mounted 25 mm gun and a 7.62 mm machine gun, can carry half a dozen infantrymen. It can also carry wire-guided antitank missiles like the Tube-launched, Optically tracked, Wire-guided (TOW) missile. The infantrymen can use automatic, rifle-caliber "firing point weapons" while the vehicle is on the move, and they can leave the rear of the machine with their regular infantry weapons. Some APCs have open tops so they can mount mortars and fire them on the move. By the Iraq War, the United States had adopted a 120 mm mortar—one similar to the very effective Communist-bloc weapon that had been roundly cursed by American troops in Korea. And the American weapon was self-propelled.

On the ground and out of their vehicles, modern American infantrymen have many weapons that would be familiar to soldiers in other wars, such as the M 16 rifle, an improved version of the weapon that appeared during the Vietnam War. In the Iraq War, they had armor vests, familiar to grunts in the last half of the Korean War, but able to stop bullets as well as shell fragments. (But also bulkier.) The TOW had been around for quite a while, too. It trailed a wire that carried directions from a soldier who guided it to a target up to 2.3 miles away. A newer and better antitank missile is the Javelin, which, after it is aimed and fired, guides itself to the target. The Javelin has another cute trick: it flies above the tank and lands on its top, where the armor is thinnest.

War in the Air

As the opening of the Gulf War demonstrated, the highest tech of the new high-tech war is found well above the surface of the earth. Helicopters, planes, long-range missiles, and even satellites play a major role in modern war. The Global Positioning System (GPS), based on satellites, now guides Tomahawk missiles instead of the ground-scanning radar system used in the Gulf War. This provides accurate guidance in spite of the weather.

Tomahawk missiles may be launched from either ships or planes. The latter, a slightly different type, is called a Conventional Air-Launched Cruise Missile (CALCM)—"conventional" because the first cruise missiles were designed to carry nuclear warheads. The CALCM is just one of many types of airplane-launched weapons in modern warfare. It is a powered missile that is used when it is desirable to hit a target a long distance from the plane. Unpowered glide bombs are released closer to the target, but are guided and just as precise. Some are guided by shining a laser beam on the target so the bomb can home in on the reflection of the beam, some have a GPS guidance system, some have built-in inertial guidance systems, and some home in on heat—especially antiaircraft and antitank weapons. Helicopters have become extremely effective in the antitank role by using rockets like the Hellfire missile. One of the more interesting devices at the current time is the unmanned aerial vehicle (UAV). UAVs are small, have tremendous range and endurance, and can carry television cameras to provide "real-time" intelligence from any battlefront. They can also carry weapons. One made headlines around the world when it was used to assassinate a suspected terrorist in Yemen.

The First Test

The world got a glimpse of all this high-tech hardware during the Gulf War, but it got its real showing in Afghanistan a decade later. On paper, Afghanistan looked like anything but a pushover. The Soviet Union, a first-class power, had invaded Afghanistan in 1979. In 1989, the Soviets left, conceding that the country was impossible to conquer. The Soviet failure in Afghanistan was one of the leading causes of the demise of the Soviet Union and the end of the Cold War.

The U.S. effort in Afghanistan was conducted almost entirely with air power and guided missiles. At first, there seemed to be a lot of bombing of barren mountains and absolutely no progress. Then a key component got involved: special forces operatives. What they did was locate targets, call in strikes, and identify the targets with lasers and other means. The United States did not have to put infantry on the ground, because they were already there, in the form of the some-what undisciplined but very tough troops of the Northern Alliance, which had been holding off the Taliban for months. Accurate bomb-ing was too much for the also very tough Taliban troops. The Taliban eventually collapsed. The Afghanistan War wasn't exactly the sort of thing Billy Mitchell had predicted, but it was as close as anyone has ever come to a victory through pure air power.

Military theorists who had been calling for a "lighter, more agile" army were encouraged, especially those who were among the civilian leadership in the Pentagon. When the next war came along, the inva-sion of Iraq, the United States used the "light and lively" approach. The war was over in three weeks. It looked as if high tech was the answer to any military problem. We had guided missiles that could pick out military targets and destroy them without leveling whole cities as in World War II. We could blind our enemies electronically. We could listen in on their telephone conversations. We could take away their electric power. We could see in the dark, locate their armies in sandstorms, and blast them from the sky while they had no idea an enemy was near. We could send armored columns through their coun-try so rapidly they would still be preparing for an attack when we'd hit their rear.

But then, it appeared that there is still a part of warfare that is dis-turbingly and stubbornly low tech.

21

Low Tech

French partisan wields a Sten submachine gun as he and a U.S. Army lieutenant engage German troops in street fighting in 1944. Partisan guerrillas played a part in World War II. Guerrillas have played much bigger roles since then, especially in Vietnam and Iraq. (U.S. Army photo, National Archives.)

Men and Machines

WHEN WE CONSIDER turning points in warfare, we tend to think in terms of weapons, which, over the ages, have become increasingly sophisticated. In the high-tech warfare we've been discussing, the key weapons are machines so sophisticated that only a few specialists really understand them. The dazzling progress that weaponry, the machines of war, has shown in the last generation can lead us to forget that wars are not made by machines, but by human beings using those machines. And this can lead us to forget something else.

Human bodies have made no such dazzling progress in the last generation or the last 1,000 generations. A poisoned arrow could kill a Scythian warrior 3,000 years ago, and a similar arrow from a primitive crossbow could kill an American soldier in modern Vietnam. And many did. Nothing could be more low tech than an excrement-smeared punji stick, but such spikes caused many casualties and some deaths in Vietnam. In Iraq, there are many far less primitive weapons. Automatic rifles and rocket launchers seem to be everywhere. There are so many caches of explosives in shells, bombs, and rockets that the military authorities have not been able to guard them all. But rifles are nobody's idea of high tech, and what the army calls "improvised explosive devices" are as low tech as anything that goes "BANG!" Although they're not high tech, the tools of death are unusually abundant in Iraq, and the killing goes on.

Halfway through 2004, the number of Americans killed in Iraq was over 900, and the vast majority of them died after May 1, 2003, when President George W. Bush said the war was over. Why does the war still go on?

This is a question of intelligence gathering that has nothing to do with the nonexistent "weapons of mass destruction" but may be more important. Traditionally, military intelligence has tried to gather information on the enemy's military forces: How many tanks, planes, and so on does he have? How serviceable are they? What about the morale of his troops? Their level of training? It has looked much less closely at the population as a whole.

We need to study nations, not just military forces. We need to know why people fight, and why they stop fighting.

In 1917, Russia was ripe for revolution. It had not recovered from a revolution only thirteen years before. People were starving, the Russian army was being beaten, with hundreds of thousands of soldiers dying for no apparent reason, and the government was run by lunatics, according to Maurice Paleologue, the French ambassador.[1]

The Russians made peace so they could change the government. In World War II, the bitterly divided and feuding French body politic was a victim of "shock and awe"—something the blitzkrieg delivered far more efficiently than the United States and its partners in the Iraq War. No sensible person in Iraq thought his country had an invincible army, but many sensible French people believed that of their army. But there was a French resistance. In 1945, there was no German resistance. Germany had been overrun, and the Germans were exhausted. German military deaths—4 million—cut deeper into that country's young manhood than the more numerous—7 million— Soviet deaths, because the Soviet Union was a larger country.[2] And what did the Germans have to fight for? Adolf Hitler, who killed himself? The Japanese might have gone on fighting, even though they had nothing with which to fight. But conditioned by generations of emperor worship, when the emperor told them to stop, they stopped.

The Korean War was halted by an armistice that was signed after several communist offensives had been bloodily repulsed in 1952 and 1953, which had convinced the Chinese and North Koreans that there was no chance at that time of uniting the peninsula. But officially, the war is still on, and North Korea is still a problem. The Vietnamese people wanted independence after generations of misrule by France.[3] The Viet Cong and the North Vietnamese government were more convincing supporters of independence than the South Vietnam government, which began to resemble an American puppet. The communist-led Vietnamese believed that they were more patient than the Americans and that if they held on long enough, they would win. They were right.

Walking in the Other Guy's Shoes

Many pundits proclaim that Americans don't understand people of other cultures, which is why our record of fighting guerrillas is hardly enviable. It probably is difficult to think like an Iraqi. But do Americans ever suppose that Iraqis think like Americans? After all, we are both human beings. We must have some feelings in common.

Right now, in 2004, with a presidential election to be held in November, President George W. Bush is not popular with Democrats. He is, in fact, loathed by them almost as much as his predecessor, Bill Clinton, was loathed by Republicans. Now suppose a foreign nation invaded the United States, saying it intended to "liberate" the American people from Clinton/Bush (choose the object of your loathing). How many Republicans do you suppose would greet the anti-Clinton liberators with "sweets and flowers"? About the same number of Democrats who would hail the anti-Bush liberators, right? There would be more bombs than "sweets and flowers."

So why is it such a surprise that instead of the "sweets and flowers" Ahmed Chalabi predicted, our troops saw only sullen crowds, then looting, then bombs?

Of course, neither Clinton nor Bush has been the sort of vicious tyrant Saddam Hussein was. Life under Saddam was difficult for anyone with political ideas different from Baathist orthodoxy. Criticizing the great leader was suicidal. But when we speak of Saddam killing thousands of his own people, we forget that Adolf Hitler killed millions of his own people without noticeably impairing his popularity. The U.S. federal and state governments have also killed hordes of their own people—people called Cherokees, Apaches, Sioux, and similar names. In the mid-nineteenth century, the state of California organized a program of genocide, in which it paid bounties for dead Indians. Many if not most of those thousands Saddam killed were Kurds, people about as popular with the Arabs as Apaches were to New Yorkers in 1880.

No question about it, Saddam is an evil man. And few Iraqis will miss him. But he is an Iraqi—one of their own—and he was removed by foreigners. Iraq, for a thousand or so years, has had trouble with foreigners. Once the center of the Arab and Muslim worlds, it has been subjugated by the Seljuk Turk, Mongol, Persian, Ottoman Turk, and British armies. This makes it difficult for the average Iraqi to believe that the Americans have come as liberators. This is especially true when the Americans say that after setting up a "sovereign" Iraqi government, they'll be keeping troops in Iraq for years to come.

Let's put ourselves in the Iraqis' place again.

Suppose the foreigners who say they have liberated you set up bases in Washington, D.C., New York, Los Angeles, Chicago, and every county seat. They surround these bases with razor wire and concrete barriers and don't let any "natives" like you get near them. They set up checkpoints on the street and shoot you if you don't stop. At 2 or 3 o'clock in the morning, they pull up in front of houses in armored vehicles, kick in doors, scream at the inhabitants in a language you can't understand, scatter the contents of drawers and closets on the floor, and, finding nothing, leave. Sometimes, they leave with male members of the family, never explaining why. Sometimes, those men come back after a few weeks; sometimes, their families are still waiting for them. But aside from the raids and the checkpoints, the "libera-tors" have nothing to do with the "liberated." Unless you try to pass their vehicles in your car. Then they shoot you.

Would you feel grateful for your liberation?

Would you feel you were better off because sewage disposal is broken down, parts of town have no water, electricity is a sometimes thing, you have to wait in miles-long lines to get fuel, and, because of criminals, you take your life in your hands if you go out after dark?

Americans haven't been thinking like Iraqis. They haven't been thinking like Iraqis who think like Americans, either. In Iraq, Ameri-can thinking has been based on a fantasy. We invade a foreign coun-try and call it "Operation Iraqi Freedom," but we inhibit all kinds of freedoms, such as movement—we even surrounded one village with barbed wire and wouldn't let residents leave or enter without permis-sion. We arrest people in Gestapo-like raids. We appoint a puppet gov-ernment composed mostly of Iraqis who have been living in exile for years or even generations. And we expect the Iraqis to think they have been freed.

So some Iraqis plant roadside bombs. Some shoot at helicopters and Humvees with RPGs. Some kill Americans and their allies with suicide bombs. American occupation authorities invariably describe these attackers as "Saddam loyalists" or "terrorists." As for their being "Saddam loyalists," the authorities never explain why, when foreign reporters question residents where an attack has taken place, the response they usually get is, "I hate Saddam, but I hate the Americans more. I wish they would all go home."

Terrorists and Guerrillas

When the authorities talk about terrorists, they ignore the difference between terrorists and guerrillas. The U.S. problem in Iraq is mostly a guerrilla problem. It's important to understand what guerrillas are, because there are probably going to be a lot of them in the future. At the present time, no other nation is in a position to challenge the United States in regular warfare. But any nation can wage guerrilla warfare if it has the motivation.

Terrorists are people who commit a violent act to draw attention to a cause. They have no hope that what they do is going to cause their adversary to change his plans in a way that will benefit their cause. The September 11, 2001, hijackers were terrorists. So was Timothy McVeigh. McVeigh also demonstrated how a terrorist or guerrilla can make a horribly effective bomb with the most humdrum materials.

The guerrilla is a soldier. He's a different kind of soldier than the regular, but, like the regular, his operations are based on strategy as well as tactics. Once upon a time, the military establishment usually thought guerrillas could only be effective if they were operating behind enemy lines while a regular army unit was facing the enemy. That's how guerrillas operated in Spain, where the word was coined, against Napoléon Bonaparte's troops. It's often a very effective strategy. It's especially effective against a highly mechanized army. It's an odd fact that the more mobile an army becomes—with some exceptions like the Central Asian horse archers—the more it depends on supply lines. In 1914, a typical infantry division required less than 100 tons of supplies a day. Much of that was fodder for the horses. In 1940, a panzer division required 300 tons a day. By the end of the war, an American armored division, much more mechanized than an early panzer division, required 1,000 to 1,500 tons a day.[4] Almost all of those supplies had to be hauled by vehicles open to attack by guerrillas.

The idea that guerrillas are only an important factor if they are working with a regular unit in the field may be caused by the regular military's sense of its own importance. Mao Tse-tung's guerrillas *turned themselves* into a regular army and conquered a country.[5] So did Ho Chi Minh's guerrillas in North Vietnam.

Another widespread misconception was that guerrillas needed a safe area in some friendly country where they could regroup and get new supplies. Fidel Castro's guerrillas couldn't cross a border to a friendly country. They made their own safe areas in Cuba, as Castro's lieutenant, Ernesto "Che" Guevara, explained in his book *Guerrilla Warfare*.[6]

Guerrilla strategy in the so-called wars of liberation follows the pattern set by Mao. Small, widely scattered units attack the enemy while bonding with the peasants in their areas. They recruit the peasants, grow stronger, and make heavier attacks. They hide among the people until they are strong enough to discourage enemy troops from raiding into their territory. Finally, they are able to field an army with such equipment as tanks and artillery, captured from the enemy, and defeat the enemy in the field. The Vietminh had reached that stage when they took the French fortress at Dien Bien Phu.

Lawrence of Arabia practiced what might be called a strategy of containment. He originally intended to cut the Turkish railroad between Mecca and Damascus, but he later decided to keep it open—barely. His idea was to keep the Turkish garrisons in Mecca and Medina but prevent them from having enough supplies to foray into the villages and tribal encampments of Arabia. Their presence in the two holy cities would be a constant drain on supplies the Turks could use elsewhere, and the garrisons would be too busy trying to defend the railroad to do anything else. Meanwhile, Lawrence sent agents to all the Arabs in the area to preach revolt.

In the Irish war of independence, the "Black and Tan War," the Irish revolutionists used a complex strategy involving violent and nonviolent resistance with a powerful propaganda campaign. The nonviolent portion was to follow the plan of Arthur Griffith's Sinn Fein (Gaelic for Ourselves Alone) Party and set up a parallel government with a legislature and courts separate from the British system. The British tried to enforce London's rule with the Royal Irish Constabulary, but the population began to shun the police, and the police began to look for other work. The British sent English intelligence officers to Dublin to arrest or kill officials of the Irish Republic. To fill the ranks of the Royal Irish Constabulary, the British sent an army of unemployed war veterans, nicknamed the Black and Tans because they wore a combination of army and police uniforms.

To counter the British intelligence agents, Michael Collins developed his own agents in Dublin Castle, the seat of British government in Ireland. They identified the British agents and on the morning of November 21, 1920, Collins's agents killed them all. At the same time, Irish Republican Army columns in the countryside were ambushing Black and Tan units. And while all this was going on, Erskine Childers, the Irish propaganda chief, was informing the British public and the world at large of Black and Tan atrocities in Ireland.[7]

The Irish situation is particularly instructive because the rebels had no armor, no artillery, hardly any machine guns, and not enough rifles to go around. They were successful against one of the greatest military powers of the time. If they had been contending against Hitler's Germany or Joseph Stalin's Soviet Union, the results would undoubtedly have been different. But the intelligent guerrilla tailors his strategy to his opponent.

In Iraq, the guerrillas began by attacking American vehicles—few American soldiers travel on foot—mostly with remote-controlled roadside bombs. It may have been partly a Mao-style recruitment technique—attacking an unpopular occupier attracts recruits—and it may also have been an effort to encourage the Americans to continue their self-imposed isolation. Now, the attacks seem more aimed at Iraqis who cooperate with the Americans. The military authorities say this is because Iraqi civilians are "softer" targets than American military. This is just silly. A remote-controlled bomb is no more dangerous to set off beside an armored car than beside a civilian bus. A suicide bomber is just as dead if he blows himself up in a crowd of civilians or in a crowd of soldiers. The guerrillas are killing civilian collaborators to discourage other civilians from cooperating with the occupation. If no civilians cooperate with the occupation, the occupation will fail.

Describing the situation of soldiers and marines in post–May 2003 Iraq, Williamson Murray and Robert H. Scales Jr. say, "Today these 'laptop warriors' are foot soldiers performing grunt tasks no different from the British army in Palestine in the 1930s and Northern Ireland in the 1970s and 1980s, or, for that matter, the Roman army in first century Judea."[8]

That's the trouble. U.S. occupation authorities have been trying to learn from the British experience in Northern Ireland and also from the Israeli Palestinian problem. But all the evidence indicates that the British military operations in Palestine and Northern Ireland merely exacerbated bad situations, and the Israeli army has certainly not brought peace to that part of the Near East. And the work of that Roman army in the first century brought on the gory Jewish War and the mass suicide of Jewish rebels on the Masada.

We have to do better than that.

Glossary of Military Terms

A glossary of all military terms would take another book. The terms listed below are widely used in the literature. They are also sometimes misused or might be confusing or unfamiliar to the average person.

Abatis: A line of felled trees with the branches sharpened and pointed toward the enemy. An obstacle widely used before the invention of barbed wire.

Admiral: The highest-ranking naval officer. In the U.S. Navy, the progression is rear admiral (two grades), vice admiral, admiral and five-star admiral. The U.S. Navy had no admiral before David Glasgow Farragut.

Army: A military unit composed of two or more corps. Also, the entire ground forces of a nation.

Army group: Several armies (first definition, above) fighting on the same front. The Russian term for this type of unit is "front." In the 1930s and 1940s, the Communist Party in the United States applied the Russian military term to various political organizations that attempted to hide their Communist connections. By ignoring a common American definition of "front," the CPUS made a ludicrous mistake.

Automatic, semiautomatic: Automatic small arms fire, eject cartridge case, fire again and continue to do so as long as the trigger is held back and ammunition remains in the magazine or belt. Semiautomatic weapons fire, eject and reload but do not fire a second shot unless the trigger is released and depressed again. Almost all of the pistols commonly called automatics are really semiautomatic.

Battalion: A military unit consisting of several companies. Similar units in the cavalry are called squadrons.

Battery: An artillery unit corresponding to an infantry company. In times past, it was also a fort housing such an artillery unit.

Blockhouse: A small fort of widely varying construction. Spanish blockhouses in the Spanish American War were often made of wood. British blockhouses in the Second Boer War were roofed cylinders composed of two layers of corrugated iron filled with fieldstone and pebbles. On American frontier forts, they were towers overlooking the walls, often constructed of squared logs.

Bow: The most important missile weapon before the invention of gunpowder. It's a curved spring with the ends connected by a bowstring. An arrow is fitted to the bowstring and drawn back. When the string is released, the bow springs back to its former position and the bowstring propels the arrow. Bows composed of one type of material (usually wood) are called self bows. Those built of layers of the same material, laminated bows, and those containing layers of different materials, composite bows. Composite bows are more flexible than self bows, therefore they can be shorter and easier to manage on horseback. **Crossbows** (*see entry*) are a special type of bow.

Breech loader: A gun that loads at the end of the barrel opposite the muzzle. Breech loading has many advantages, especially in small arms. Breech loading makes concealment easier and makes repeating rifles, pistols and shotguns possible.

Brigade: A military unit larger than a regiment and smaller than a division.

Bunker: Something like a blockhouse—a small fortified position. Bunkers on World War II's Maginot Line and Westwall were made of reinforced concrete. In the Pacific islands during the same war, they were partially buried and roofed with palm logs, coral rock and sandbags. In the Korean War, they were holes with the above ground portion walled with sandbags and roofed with logs and sandbags.

Captain: In the U.S. Army, a captain ranks higher than a second or first lieutenant and lower than a major. The rank is equivalent to a navy lieutenant. In the U.S. Navy, captain is equivalent to an army colonel. But in the U.S. Navy, any commissioned officer commanding a ship is referred to as that ship's captain.

Cavalry, mounted infantry: Cavalry ride horses and fight mounted. Mounted infantry ride horses but dismount to fight. During the American Civil War, most cavalry outfits actually acted as mounted infantry.

Cohort: Although it's often misused as a synonym for colleague, a cohort is really 600 Roman infantrymen, the largest unit in the legion as reorganized by Marius.

Company: A military unit consisting of several platoons and smaller than a battalion.

Cone of fire: The space covered by a machine gun burst. No machine gun puts one bullet directly behind another.

Corps: A military unit composed of two or more divisions.

CPO: Chief petty officer, the highest enlisted rank in the U.S. Navy. The CPO wears a different uniform from other sailors and has his own mess.

Crossbow: A bow mounted on a stock. The bowstring is drawn back and locked in position, after which an arrow is placed in front of it. Crossbows can be drawn using both hands or using a mechanical device like a lever or a windlass, therefore they can have a much heavier draw weight than an ordinary bow.

Cuirass: Armor covering the upper torso. The first were apparently made of leather, as the name comes from the French word for leather.

Division: Military unit larger than a brigade, composed of two or more brigades or regiments.

Dugout: Underground chamber, much used during World War I for living quarters, command posts and communications centers.

Enfilade: Small arms or artillery fire on an enemy's flank. Such fire is particularly deadly because a line of men makes a much denser target when fired on from one end than from the front.

Escort carrier: Small aircraft carrier used to escort convoys in World War II.

Field marshal: European army rank equivalent to a U.S. five-star general.

Fission bomb: A nuclear bomb that releases energy by splitting uranium or plutonium atoms.

Flintlock: A gun fired by sparks, created by a piece of flint striking a piece of steel attached to the cover of a "flash pan." The sparks ignite the powder in the flash pan, which then ignites the main powder charge.

Free company: A band of mercenaries who fought for anyone who paid them. When mercenaries turned into regulars, the terms "company" and the company's leader, a "captain," were retained for one of the smaller units of a regular army.

Frigate: In the days of sail, a ship smaller than a line-of-battle-ship that was often used in commerce raids. It was the equivalent of a cruiser. Now, it's a ship smaller than a destroyer often used as a convoy escort vessel.

Fusion bomb: A nuclear bomb that combines two elements (deuterium and tritium) that form another element, helium 4, and release energy.

General: The highest army rank. In the U.S. Army there are brigadier general (one star), major general (two stars), lieutenant general (three stars), general (four stars) and general of the army (five stars). There are no five-star generals at present.

Gun: In general, any weapon that fires a projectile by means of an explosion. In artillery, a relatively long-barreled cannon that fires a projectile on a flat trajectory at relatively high velocity.

Howitzer: An artillery piece with a shorter barrel than a gun that usually fires at a higher angle of elevation.

Legion: A Roman military formation that was the equivalent of a modern division. In the eighteenth century, mixed infantry and cavalry formations were called legions.

Light infantry: In the eighteenth century, infantry troops with lightened equipment who often fought in small units and were trained in individual marksmanship and expected to think for themselves. Today, they are infantry units without as much heavy armor and artillery as other divisions and are often airborne.

Line-of-battle-ship: The largest naval vessel in the age of sail. It is the direct ancestor of the battleship, which gets its name from the big sailing ship.

Maniple: Named from the Latin word for handful. It was the smallest unit of the early Roman legion and was composed of two centuries (which in spite of the name usually did not contain 100 soldiers).

Marshal: In France, not a military rank but an honorific title.

Matchlock: A gun that was fired by a mechanism that lowered a smoldering cord (called a match) into a flash pan full of gunpowder.

Mercenary: A soldier who would fight for anyone who paid his price. Usually a member of a free company.

Militia: In U.S. law, all male residents between seventeen and forty-five years of age are members of the militia. These men are called the unorganized militia. The organized militia is composed of the National Guard. The militia referred to in the U.S. Constitution is the unorganized militia. The National Guard did not exist until early in the twentieth century.

Mortar: A short-barreled artillery piece that fires at a higher elevation than forty-five degrees. Most mortars have always been muzzle-loaders.

Muzzle loader: A gun, whether small arms or artillery, that is loaded through the muzzle.

Neutron bomb: A nuclear bomb that has a relatively low blast effect but produces an abundance of free neutrons. The Pentagon term is "enhanced radiation device." It is supposed to eliminate every living thing in a wide area but leave property largely undisturbed. It has never been used in war or even fully tested.

Non-commissioned officer (NCO): Military leaders who have enlisted for short hitches and do not have commissions. The highest NCO ranks below the lowest commissioned officer.

Percussion lock: A rifle, musket, shotgun or pistol fired by the explosion caused by a blow on a priming mixture (usually fulminate of mercury) in the primer cap that is not part of the metallic cartridge.

Petty officer: Another name for an NCO in the navy or coast guard.

Pilum: The Roman throwing spear. It had a small steel head on a long metal shaft that was attached to a heavy wooden shaft. The metal shaft prevented an enemy from cutting off the spear if it stuck in his shield, and the metal shaft would either bend or hinge down from the wooden shaft when it hit something solid. That made it impossible for the enemy to throw it back at the Romans.

Pistol: A small gun designed to be used in one hand. It was an important weapon for horse cavalry, but has since mainly been used for personal protection by officers and specialists like machine gunners and artillery men. Some people think pistols and revolvers are different things. They aren't. Revolver is a type of action. In the nineteenth century there were rifles that were also revolvers.

Platoon: A military unit consisting of several squads and smaller than a company.

Poison gases: In World War II, lethal poison gases (excluding such agents as tear gases) were in two groups—choking gases like phosgene and chlorine, and blister gases like mustard gas and lewisite. Both were deadly, but the blister gases were more fearsome. They would burn any exposed tissue, from skin to lungs, and people died from even the external burns. After World War I, scientists began developing nerve gases, which can penetrate the skin and destroy the body's nervous systems. They can destroy the respiratory system, the digestive system, muscular control and the heart. They can kill in a matter of minutes.

Privateer: A privateer was a legal pirate, but legal only as long as he preyed on the ships of the enemy of the country that issued his "letters of marque and reprisal."

Regiment: A military unit that includes several battalions and is smaller than a brigade.

Repeating rifle: A rifle that holds cartridges in a magazine and can be reloaded by a simple motion like opening and closing a bolt, pulling down and lifting a lever or moving a forend back and forth. Semiautomatic rifles that require only another pull of the trigger are also repeaters, but are most often called semiautomatics.

Rifle: An individual firearm with a rifled barrel that is usually held in two hands and braced against the shoulder. In the navy, the term also meant the huge cannons on warships if they were rifled.

Rifling: Grooves cut into the interior of a gun barrel to make the projectile spin, which increased its accuracy and stability.

Rocket propelled grenade: Actually, there's no such animal. Someone in Vietnam thought that the RPG in RPG 7, a Russian-invented antitank weapon stood for rocket propelled grenade. It doesn't. RPG appears on a variety of Russian antitank weapons, such as the RPG 43, an antitank grenade, and the RPG 2, a recoilless gun developed from the German panzerfaust. The RPG 7 is a recoilless gun that shoots a rocket-assisted shell far enough ahead of the gunner to keep him from being incinerated by the rocket engine when it ignites. There's no quick way to describe this weapon, so the term rocket propelled grenade—stupid as it is—has caught on and will be with us for as long as RPG 7s and their knock-offs are around.

Sapper: British term for a military engineer: sappers were people who dug saps (trenches leading to the walls of a besieged city). Somebody in Vietnam heard that North Vietnam had troops called sappers and leaped to the conclusion that the word meant special shock troops. A lot of confusion came from public information offices in Vietnam.

Scutage: In the middle ages, kings charged knights scutage—the shield fee (from *scutum*, Latin for shield)—in lieu of military service. With that money, the king could hire mercenaries, who were more efficient warriors and had no feudal limits on their length of service.

Shrapnel: The invention of Lieutenant (later General) Henry Shrapnel. Shrapnel's invention was a hollow shell filled with gunpowder and lead balls. It was an antipersonnel weapon for use at ranges beyond the reach of grape shot. In a modernized form it was heavily used in World War I. Though people today are reported to be wounded by shrapnel, their wounds are really caused by shell or bomb fragments. The last reported use of shrapnel was in the Spanish Civil War of 1936–1939.

Smoothbore: A gun, whether an individual weapon like a musket or a cannon, without rifling.

Spear: A pointed weapon on a relatively long shaft. Spears take many forms, from the short, light javelin, to the long lance of the horseman, to the ponderous pike of ancient Macedon or renaissance Europe. In a way, the rifle-mounted bayonet is a kind of spear.

Squad: The smallest unit of infantry in the U.S. Army. It is usually led by a sergeant. In the British army, it's called a section; in the German, a gruppe.

Squadron: What they call a battalion in the cavalry. The term is also used for naval and air force units.

Theater: Term used by the Allies in World War II to define geographical areas of operations, e.g. European Theater, Mediterranean Theater.

Troop: The cavalry equivalent of an infantry company. At various times, the cavalry called a unit this size a company, too.

Weapons of mass destruction: A political term for nuclear weapons, chemical weapons and biological weapons. All of them can destroy human life, although not always in mass. In World War I, when it was most used, poison gas (chemical) fell far short of mass destruction. It could not be compared with napalm (a non-WMD) as used in World War II. Neither gas, nor disease, nor neutron bombs can destroy much other than life.

Wheellock: A gun that is fired by a spinning wheel rubbing a piece of iron pyrites. There is no noticeable delay between pulling the trigger and the discharge of the gun, as there is with the flintlock. However, the wheellock mechanism is more expensive and delicate than that of the flintlock, so it never replaced the matchlock, as the flintlock did. But in the sixteenth century, before the flintlock appeared, the wheellock pistol was a basic cavalry arm. Matchlock pistols would have been awkward and dangerous to use on horseback.

Notes

1 The Basics: Point and Edge, Arrows and Armor

1. Oakley, Kenneth P. *Man the Toolmaker*. Chicago: University of Chicago Press, 1957, pp. 23–25.

2. Ibid., pp. 38–54. Piggott, Stuart. *The Dawn of Civilization*. New York: McGraw-Hill, 1961, pp. 36–37.

3. Derry, T. K., and Trevor Williams. *A Short History of Technology: From the Earliest Times to A.D. 1900*. New York: Oxford University Press, 1961, pp. 114–15.

4. Diaz, Bernal de Castillo. *The Bernal Diaz Chronicles*. Garden City, NY: Doubleday, 1956, p. 231; Prescott, William H. *The Conquest of Mexico*. Garden City, NY: Blue Ribbon, 1943, p. 257.

5. Pope, Saxton T. *Bows and Arrows*. Berkeley: University of California Press, 1962, pp. 54–55.

6. McEvedy, Colin. *The Penguin Atlas of Ancient History*. New York: Viking Penguin, 1988, pp. 54–55.

7. Chadwick, John. *The Mycenaean World*. Cambridge, UK: Cambridge University Press, 1976, pp. 69–83; Palmer, Leonard R. *Mycenaeans and Minoans*. New York: Knopf, 1962, pp. 91–119.

8. Derry and Williams, *Short History of Technology*, pp. 121–22.

9. Yumoto, John M. *The Samurai Sword: A Handbook*. Rutland, VT: Tuttle, pp. 97–108; Harris, Victor, ed. *Swords and Hilt Weapons*. New York: Barnes and Noble, 1993, pp. 148–51.

10. Harris, Victor, ed. *Swords and Hilt Weapons*. New York: Barnes and Noble, 1993, pp. 18, 31, 33; Oakeshott, R. Ewart. *The Archaeology of Weapons*. New York: Praeger, 1960, pp. 86–87.

11. Derry and Williams, *Short History of Technology*, p. 126.

12. McNeill, William H. *The Pursuit of Power*. Chicago: University of Chicago Press, 1982, p. 26.

13. Nixon, Ivor Gray. *The Rise of the Dorians*. New York: Praeger, 1968, p. 42.

14. Ibid., pp. 17–18; Piggott, *Dawn of Civilization*, pp. 154–55.

15. Nixon, *Rise of the Dorians*.

16. Prescott, William H. *The Conquest of Mexico*. Garden City, NY: Blue Ribbon, 1943, p. 291.

17. Payne-Gallwey, Sir Ralph. *The Crossbow*. London: Holland, 1986, pp. 27–29.

18. Tarassuk, Leonid, and Claude Blair, eds. *The Complete Encyclopedia of Arms and Weapons*. New York: Bonanza, 1986, p. 96.

19. Pope, *Bows and Arrows*, p. 31; Payne-Gallwey, *Crossbow*, p. 22.

20. Prescott, *Conquest of Mexico*.

21. Piggott, *Dawn of Civilization*, pp. 74–75.

2 Urbanite Warfare

1. Each of the ten generals in the Greek army rotated command. On this day, it was Miltiades' turn.

2. Herodotus, Aubrey de Selincourt, trans. *The Histories*. Baltimore, MD: Penguin, 1960, book 6, p. 394.

3. Rodgers, William L. *Greek and Roman Naval Warfare*. Anapolis, MD: U.S. Naval Institute, 1964, p. 16.

4. Liddel Hart, Basil H. *Strategy*. New York: Praeger, 1960, pp. 27–28; Weir, William. *50 Battles That Changed the World*. Franklin Lakes, NJ: Career/New Page, 2001, pp. 12–13; Herodotus, *The Histories*, p. 404.

5. Snodgrass, A. M. *Arms and Armour of the Greeks*. Ithaca, NY: Cornell University Press, 1967, pp. 90–91.

6. Ibid., p. 53.

7. Piggott, *Dawn of Civilization*, pp. 74–75.

8. Dunan, Marcel, ed. *Larousse Encyclopedia of Ancient and Medieval History*. New York: Harper and Row, 1963, p. 51.

9. Palmer, *Mycenaean and Minoans*. pp. 172–74; Chadwick, *Mycenaean World*, pp. 164–71.

10. Snodgrass, *Arms of Armour of the Greeks*, p. 24

11. Adcock, F. E. *The Greek and Macedonian Art of War*. Berkeley: University of California Press, 1957, pp. 4–6; Morrison, Sean. *Armor*.

New York: Thomas Y. Crowell, 1963, pp. 39–40; Snodgrass, *Arms and Armour of the Greeks*, p. 49

12. Adcock, *Greek and Macedonian Art of War*, pp. 14–28.

13. Marsden, E. W. *Greek and Roman Artillery: Historical Development*. Oxford: Oxford University Press, 1999, pp. 58–60.

14. Adcock, *Greek and Macedonian Art of War*, pp. 26–28; Snodgrass, *Arms and Armour of the Greeks*, pp. 114–30; Montross, Lynn. *War through the Ages*. New York: Harper and Row, 1960, pp. 16–23.

15. Morrison, *Armour*, pp. 51–55; Montross, *War through the Ages*, pp. 43–51.

16. Morrison, *Armour*, pp. 53–72; Montross, *War through the Ages*, pp. 46–48; Simkins, Michael. *Warriors of Rome: An Illustrated Military History of the Roman Legions*. London: Blandford, 1988, pp. 28–30.

3 The Rulers of the Plains

1. McEvedy, Colin. *The Penguin Atlas of Ancient History*, New York: Viking Penguin, 1988, 24–89.

2. Plutarch. *Lives of the Greeks and Romans*. "The Life of Crassus" covers Crassus's ill-fated campaign in detail. Chicago: University of Chicago, 1962.

3. Parker, Geoffrey. *Cambridge Illustrated History of War*. New York: Cambridge University Press, 1995, p. 9.

4. Ibid., p. 5.

5. Ibid., p. 3

6. Piggott, *The Dawn of Civilization*, pp. 74–75.

7. Ibid., pp. 322–23.

8. Gurney, O. R. *The Hittites*. Baltimore, MD: Penguin, 1969, pp. 104–05.

9. Ibid., p. 106.

10. The stirrup is often said to have been introduced in western Europe in the eighth century, but there is evidence that it is much older. Chinese records show that the stirrup, developed by Central Asian nomads, was adopted in the Middle Kingdom by the Northern Wei dynasty around 477. (See Brownstone, David, and Irene Franck. *Timelines of War*. Boston: Little, Brown, 1994, p. 91.) R. Ewart Oakeshott, in *The Archaeology of Weapons* (New

York: Praeger, 1960, p. 85), points out that Central Asian sculptures show riders using stirrup loops as early as the second century B.C. At any rate, it's certain that the Huns and the Alans had them during the last years of the Roman Empire and that the Goths learned to use them from the nomads.

11. Konstam, Angus. *Historical Atlas of the Celtic World*. New York: Facts on File, 2001, p. 16.

12. Bury, J. B. *History of the Later Roman Empire: From the Death of Theodosius I to the Death of Justinian*. New York: Dover, 1958, vol. 1, pp. 241–43.

13. Ibid., pp. 292–94; also see Weir, William. *Fatal Victories*. New York: Avon, 1993, pp. 15–27; Gordon, C. D. *The Age of Attila: Fifth-Century Byzantium and the Barbarians*. Ann Arbor: University of Michigan Press, 1966, pp. 106–8.

4 Stonewalling the Nomads

1. Prawdin, Michael. *The Mongol Empire: Its Rise and Legacy*. New York: The Free Press, 1961, p. 212.

2. McNeill, William H. *Plagues and Peoples*. New York: Random House, 1998, pp. 161–207.

3. Lamb, Harold. *Genghis Khan, Emperor of All Men*. New York: Bantam, 1953, p. 84.

4. Weir, *50 Battles That Changed the World*, p. 74

5. Heany, Seamus. *Beowulf*. New York: Farrar, Straus, and Giroux, 2000.

6. Pratt, Fletcher. *The Battles That Changed History*. Garden City, NY: Doubleday, 1956, p. 53.

7. Brice, Martin. *Forts and Fortresses*. New York: Facts on File, 1990, pp. 38–41.

8. Van Creveld, Martin. *Technology and War: From 2000 B.C. to the Present*. New York: The Free Press, 1989, p. 31.

9. Brown, R. Allen, with Michael Prestwich and Charles Coulson. *Castles: A History and Guide*. Poole, UK: Blandford, 1980, p. 12.

10. Ibid., p. 12.

5 Private Armies and Public Policy

1. McNeill, *Pursuit of Power*, p. 64.
2. Ibid.
3. Grossman, Dave. *On Killing: The Psychological Cost of Learning to Kill in War and Society*. Boston: Little, Brown, 1995, pp. 249–90.
4. McNeill, *Pursuit of Power*, p. 66
5. For information on some of these medieval militias, see Payne-Gallwey, *Crossbow*, pp. 223–36.
6. For the composition of medieval armies, see Brown, Prestwich, and Coulson, *Castles*, pp. 69–73.
7. "Catalan Company." Wikipedia, the Free Encyclopedia, http://en2.wikipedia.org/wiki/Catalan_Grand_Company.
8. McNeill, *Pursuit of Power*, p. 74.
9. Ibid., p. 75.
10. Perrin, Noel. *Giving Up the Gun: Japan's Reversion to the Sword, 1543–1879*. Boston: David R. Godine, 1979.
11. Payne-Gallwey, *Crossbow*, pp. 90–91; Tarassuk, Leonid, and Claude Blair, eds. *The Complete Encyclopedia of Arms and Weapons*. New York: Bonanza, 1982, pp. 143–47.

6 The Devil's Snuff: The Gunpowder Revolution

1. Perrin, *Giving Up the Gun*, pp. 14–19
2. Weir, *Fatal Victories*, p. 121.
3. Kure, Mitsuo. *Samurai: An Illustrated History*, Rutland, VT: Tuttle, 2001, pp. 78–79.
4. Ibid., p. 79.
5. McNeill, *Pursuit of Power*, p. 39.
6. Van Crevald, *Technology and War*, p. 83; McNeill, *Pursuit of Power*, p. 39.
7. Levathes, Louise. *When China Ruled the Seas*. New York: Simon and Schuster, 1994, pp. 50–54.
8. Perrin, *Giving Up the Gun*, p. 25.
9. Ibid., p. 27.

Notes

10. Stone, George Cameron. *A Glossary of the Construction, Decoration, and Use of Arms and Armor in All Countries and in All Times*. New York: Jack Brussell, 1961, p. 269.

11. Smith, W. H. B. *Small Arms of the World: A Basic Manual of Military Small Arms*. Harrisburg, PA: Stackpole, 1960, pp. 4–5.

12. Peterson, Harold L. *The Treasury of the Gun*. New York: Golden, 1962, p. 32.

13. McNeill, *Pursuit of Power*, p. 81.

14. Peterson, *Treasury of the Gun*, p. 33.

15. Ibid., p. 35

16. Jobe, Joseph, ed. *Guns: An Illustratred History of Artillery*. Greenwich, CT: New York Graphic Society, 1971, p. 12.

17. Peterson, *Treasury of the Gun*, p. 35.

18. Smith, *Small Arms of the World*, p. 7.

19. Weir, *50 Battles That Changed the World*, pp. 140–49.

20. McNeill, *Pursuit of Power*, pp. 83–87; Jobe, *Guns*, pp. 18–23.

21. Duffy, Christopher. *Siege Warfare: The Fortress in the Early Modern World, 1494–1660*. New York: Barnes and Noble, 1996, pp. 1–42. Duffy presents a truly comprehensive outline of the development of the bastioned fortress.

22. Pope, Dudley. *Guns*. London: Hamlyn, 1965, p. 67; Jobe, *Guns*, p. 38.

23. Duffy, *Siege Warfare*, p. 19

7 Command of the Sea: War Under Sail

1. Derry and Williams, *Short History of Technology*, p. 196.

2. Ibid., p. 200.

3. Ibid.

4. Ibid., p. 201.

5. McNeill, *Pursuit of Power*, p. 70.

6. Derry and Williams, *Short History of Technology*, p. 204.

7. Cipolla, Carlo M. *Guns, Sails, and Empire*. New York: Minerva, 1965, pp. 80–82.

8. Levathes, Louise. *When China Ruled the Seas*. New York: Simon and Schuster, 1994, details the little-known period when China appeared ready to conquer and colonize half the world.

9. Weir, *50 Battles That Changed the World*, p. 39.

10. Indians and other Asians adopted the Arab term. "Firangi" became a synonym for Europeans in India and with slight distortions in China and Japan. It was nicer than the Chinese and Japanese "Nam Ban," meaning "southern Barbarian." European sailors had to approach China and Japan from the south.

11. Weir, *50 Battles That Changed the World*, p. 36.

12. Ibid., p. 37.

13. Ibid., p. 40.

8 The Bloody Birth of Standing Armies

1. McNeill, *Pursuit of Power*, pp. 105–14

2. Ibid.

3. For a detailed account of those wars, see Weir, Alison. *The Wars of the Roses*. New York: Ballantine, 1995.

4. Cruickshank, Charles G. *Elizabeth's Army*. Oxford: Clarendon, 1966, p. 55.

5. McNeill, *Pursuit of Power*, pp. 102–04.

6. Fuller, J. F. C. *A Military History of the Western World*. New York: Da Capo, 1955. Vol. 2, p. 21.

7. Ibid., p. 557.

8. Ibid.

9. Ibid.

10. Duffy, Christopher. *Siege Warfare*. New York: Barnes & Noble, 1979, pp. 58–105, gives details of the intricacies of fortification and siegecraft in the Eighty Years' War.

11. Fuller, *Military History of the Western World*. Vol. 1, p. 558.

12. McNeill, *Pursuit of Power*, p. 130.

13. Ibid., p. 131. McNeill also says (note, p. 131), "My remarks are derived from personal experience—and surprise at my own response to drill during World War II." My response to drill during the Korean War is quite different. I thought close-order drill was a crashing waste of time, on a par with the eternal polishing of brass and shoe leather. It was certainly valuable during the American Revolution, but today the time would be much

better spent on marksmanship with the wide variety of weapons infantry must handle, squad tactics, and such skills as laying and removing land mines.

14. The most common estimate is that 8 million people perished in the Thirty Years' War, 7 million of them civilians. Other estimates range from one-third to three-quarters of the population of Germany.

15. Fuller, *Military History of the Western World*, Vol. 2, p. 45.

16. Most English historians in the seventeenth century were Protestants. Their bias lives on in the writings of their successors. As to the Germans, they were there. And their successors, even the Protestants, do not regret that they never became a Swedish colony.

17. Liddell Hart, Basil H. *Great Captains Unveiled*. Novato, CA: Presidio, 1990, p. 171.

18. Wedgwood, C. V. *The Thirty Years War*. Garden City, NY: Doubleday, 1961, p. 322.

19. Preston, Richard A., Sidney F. Wise, and Herman O. Werner. *Men in Arms: A History of Warfare and Its Relationships with Western Society*. New York: Praeger, 1962, p. 108.

20. Liddell Hart, *Great Captains Unveiled*, pp. 117–19.

21. Montross, Lynn. *War Through the Ages*. New York: Harper and Row, 1960, p. 267.

22. Wedgwood, *Thirty Years War*, p. 266.

23. Ibid., p. 321.

24. Ibid., p. 322.

25. Ibid., p. 302.

26. Liddell Hart, *Great Captains Unveiled*, p. 193; Chandler, David. *The Art of Warfare on Land*. London: Hamlyn, 1974, p. 122.

27. Chandler, *Art of Warfare on Land*, p. 122.

28. Pratt, *Battles That Changed History*, p 178.

29. For a detailed description of the military use of preindustrial infrastructure, see van Creveld, *Technology and War*, pp. 37–49.

9 Quiet Revolution: The Age of Limited Warfare

1. Preston, Wise, and Werner, *Men in Arms*, p. 136.

2. Wilkenson, Frederick. *Edged Weapons*. Garden City, NY: Double-day, 1970, pp. 164–68.

3. Van Creveld, *Technology and War*, p. 95.

4. McNeill, *Pursuit of Power*, pp. 138–40.

5. Peterson, Harold L. *The Book of The Continental Soldier*. Harris-burg, PA: Stackpole, 1968, p. 26.

6. Peterson, Harold L. *Arms and Armor in Colonial America*. New York: Bramhall House, 1956, p. 163.

7. Preston, Wise, and Werner, *Men in Arms*, p. 137.

8. Ibid., p. 143.

9. Ibid., p. 144.

10. Fuller, *Military History of the Western World*, vol. 2, p. 346.

11. Steuben was a good drill master, but he was also a good self-promoter. He was neither a baron nor a general in Prussia, but a captain retired on half pay.

12. McNeill, *Pursuit of Power*, pp. 170–74.

10 Not Armies, but Nations

1. Bernier, Olivier. *Louis XIV: A Royal Life*. Garden City, NY: Dou-bleday, p. 318

2. Wolf, John B. *Louis XIV*. New York: Norton, 1974, p. 565; see also Weir, *Fatal Victories*, pp. 81–98.

3. McNeill, *Pursuit of Power*, p. 190.

4. Fuller, *Military History of the Western World*, vol. 2, p. 342.

5. Weir, *50 Battles That Changed the World*, p. 62.

6. McNeill, *Pursuit of Power*, p. 192.

7. Delbruck, Hans. *The Dawn of Modern Warfare. Vol. 4, History of the Art of War*. Lincoln: University of Nebraska Press, 1990, p. 396.

8. Ibid.

11 The American Civil War

1. Montross, Lynn. *War Through the Ages*. New York: Harper & Row, 1960, p. 633.
2. Millis, Walter. *Arms and Men: A Study of American Military History*. New York: Mentor, 1958, p. 109.
3. McNeill, *Pursuit of Power*, pp. 233–35.
4. Millis, *Arms and Men*, p. 79.
5. McPherson, James M. *Battle Cry of Freedom*. New York: Ballantine, 1989, pp. 279–80.
6. Ibid., p. 373.
7. Ibid., p. 375.
8. Ibid., p. 376.
9. Millis, *Arms and Men*, p. 80.
10. The Diagram Group. *Weapons: An International Encyclopedia from 5000 BC to 2000 AD*. New York: St. Martin's, 1990, p. 174.
11. Leckie, Robert. *The Wars of America*. Edison, NJ: Castle, 1998, p. 424.
12. McPherson, *Battle Cry of Freedom*, p. 377.
13. Ibid.
14. Peterson, *Arms and Armor in Colonial America*, p. 163.
15. Keegan, John. *A History of Warfare*. New York: Random House, 1993, pp. 360–61.
16. Weir, *Fatal Victories*, p. 146.
17. Keegan, *History of Warfare*, p. 305.
18. Weir, *Fatal Victories*, p. 143.

12 War by the Timetable

1. For details of the Serbian plot, see Weir, *Fatal Victories*, pp. 166–77.
2. Tuchman, Barbara. *The Guns of August*. New York: Dell, 1963, p. 99.
3. Van Creveld, *Technology and War*, p. 158.
4. Ibid., p. 159.
5. Keegan, *History of Warfare*, p. 307.
6. McNeill, *Pursuit of Power*, p. 249.

13 Out of Africa

1. Ransford, Oliver. *The Battle of Majuba Hill: The First Boer War*. New York: Crowell, 1968, pp. 26–27; Weir, *Fatal Victories*, p. 156.

2. Ransford, *Battle of Majuba Hill*, p. 27.

3. Ibid.

4. Reitz, Deneys. *Commando: A Boer Journal of the Boer War*. London: Faber and Faber, 1929, p. 21–22.

5. Smith, W. H. B., and Joseph Smith. *Small Arms of the World: A Basic Manual of Military Small Arms*. Harrisburg, PA: Stackpole, 1960, p. 43.

6. Ransford, *Battle of Majuba Hill*, p. 50.

7. Ibid., p. 88.

8. Carter, Thomas Fortescue. *A Narrative of the Boer War*. Capetown: Juta, 1896, p. 276.

9. Davis, Richard Harding. *Notes of a War Correspondent*. New York: Scribners, 1911, p. 142.

10. Pakenham, Thomas. *The Boer War*. New York: Random House, 1979, p. 141.

11. Weir, William. *Soldiers in the Shadows: Unknown Warriors Who Changed the Course of History*. Franklin Lakes, NJ: New Page, p. 147.

14 Battlewagons

1. The British navy was using monitors as late as World War I. In 1915, it sent two monitors into the Rufiji Delta in German East Africa to destroy the German cruiser *Konigsberg*. Weir, *Soldiers in the Shadows*, pp. 129–30.

2. See Mahan's most influential book, *The Influence of Sea Power upon History*. Williamstown, MA: Corner House, 1978.

3. Bain, David Haward. *Sitting in Darkness*. Boston: Houghton Mifflin, 1984, p. 56.

4. Miller, Tom. "Remember the Maine." *Smithsonian* (February 1998); Allen, Thomas. "Remember the Maine?" *National Geographic* (February 1998).

5. O'Toole, G. J. A. *The Spanish War: An American Epic*. New York: Norton, 1984, p. 176

6. Ibid., p. 192.

7. Ibid.

8. Ibid., pp. 365–66.

15 The Meat Grinder

1. Ellis, John. *The Social History of the Machine Gun*. New York: Pantheon, 1975, p. 135.

2. Keegan, John. *The First World War*. New York: Knopf, 1999, p. 295.

3. Keegan, John. *The Illustrated Face of Battle*. New York: Viking, 1988, p. 248.

4. Ibid., p. 423.

5. The one exception is the Russian front in World War II. Germany lost 4 million soldiers in World War II, most of them on the Russian front. The Soviet Union lost a mind-boggling total of 7 million men in battle. See Keegan, John. *The Second World War*. New York: Penguin, 1990, p. 590.

6. Liddell Hart, Basil H. *The Real War: 1914 to 1918*. Boston: Atlantic Monthly, 1930, p. 69.

7. Weir, William. "The War's Biggest Guns." *World War II* (May 1989).

8. Weir, *50 Battles That Changed the World*, p. 224.

9. Gudmundsson, Bruce I. *Stormtroop Tactics: Innovation in the German Army, 1914–1918*. New York: Praeger, 1989, p. 1

10. Ibid., pp. 7–10

11. Keegan, *First World War*, p. 99.

12. Ibid., pp. 132–33.

13. Marshall, S. L. A. *The American Heritage History of World War I*. New York: American Heritage, 1964, p. 77.

14. Keegan, *First World War*, p. 132.

15. These judgments are based on my firing thousands of rounds with each type of rifle.

16. Marshall, *World War I*, p. 136.

17. Diagram Group, *Weapons*, p. 16.

18. The growing pains of the tank are detailed in Liddell Hart, *Real War*, pp. 250–60. Liddell Hart was a pioneer in the development of tank warfare.

19. Keegan, *First World War*, p. 410.
20. Marshall, *World War I*, p. 312; Liddell Hart, *Real War*, p. 430; Keegan, *First World War*, p. 412.
21. Marshall, *World War I*, p. 296.

16 War Beneath the Waves

1. Keegan, *First World War*, p. 372.
2. Hoyt, Edwin P. "Predator beyond All Rules." *Military History* (February 1985).
3. Friend, Pat. "John Holland–Submarine Inventor." http://allaboutirish.com/library/people/holland.shtm; "John Holland, Father of the Modern Submarine." *Undersea Warfare* (Summer 2003). www.chinfo.navy.mil/navpalib/cno/n87/usw/issue_19/holland2.htm.
4. "Simon Lake: Biographical Sketch." www.simonlake.com/html/simon_lake/html
5. Morris, Eric, Curt Johnson, Christopher Chant, and H. P. Willmott. *Weapons and Warfare of the Twentieth Century*. Secaucus, NJ: Derbibooks, 1976, p. 75.
6. Hoyt, "Predator beyond All Rules."
7. Marshall, *World War I*, p. 204.
8. Ibid.
9. Ibid., p. 206.
10. Morris et al., *Weapons and Warfare*, p. 201.
11. Keegan, John. *Second World War*. New York: Penguin, 1989, pp. 105–10.
12. Ibid., p. 495.
13. Singh, Simon. *The Code Book*. New York: Random House, 1999, p. 184.
14. Keegan, *Second World War*, p. 129.
15. Morison, Samuel Eliot. *The Two-Ocean War*. Boston: Little, Brown, 1963, pp. 371–72.
16. Ibid., p. 496.
17. Ibid., p. 439.
18. Ibid., p. 511.

17 Blitzkrieg and Antiblitz

1. Horne, Alistair. *To Lose a Battle: France 1940*. Boston: Little, Brown, 1969, pp. 46–47.

2. Not the so-called storm troopers of the Nazi Schutzstaffel and Sturmabteilung, but the kind of storm troopers who led Erich Ludendorff's 1918 offensive.

3. For information on German machine guns and submachine guns, see Smith, *Small Arms of the World*, pp. 439–64.

4. Fuller, *Military History of the Western World*. Vol. 3, p. 341.

5. Morris et al. *Weapons and Warfare*, pp. 184–85.

6. Messenger, Charles. *The Blitzkrieg Story*. New York: Scribner's, 1976, p. 96.

7. During the purge in 1937 and 1938, Stalin executed 3 of his 5 marshals, 13 out of 15 army commanders, 110 out of 195 division commanders, and 186 out of 406 brigadier generals.

8. Brice, Martin. *Forts and Fortresses*. New York: Facts on File, 1990, p. 153.

9. Hogg, Ian V. *The Guns, 1939–45*. New York: Ballantine 1970, p. 159.

10. Diagram Group, *Weapons*, p. 91.

18 The New Queen of the Seas

1. Prange, Gordon W. *At Dawn We Slept: The Untold Story of Pearl Harbor*. New York: McGraw-Hill, 1981, p. 506.

2. Prange, Gordon W. *December 7, 1941: The Day the Japanese Attacked Pearl Harbor*. New York: Wings, 1991, pp. 108–9.

3. Prange, *At Dawn We Slept*, p. 40.

4. Ibid, p. 514.

5. Frequently Asked Questions. "Ship's Cook Third Class Doris Miller." USN, Naval Historical Center Home Page, www.history.navy.mil/faqs/faq57-4.htm.

6. Prange, *At Dawn We Slept*, pp. 532–33.

7. Layton, Edwin T. *And I was There: Pearl Harbor and Midway—Breaking the Secrets*. New York: Morrow, 1985, p. 314.

8. Ibid., p. 323; Prange, *At Dawn We Slept*, p. 558.

9. For more on why Japan allied itself with Germany, see Weir, *Fatal Victories*, p. 202

10. Prange, *At Dawn We Slept*, p. 129.

11. Morris et al. *Weapons and Warfare*, pp. 306–07.

19 War Against the Home Front

1. Messenger, *Blitzkrieg Story*, p. 34.

2. Keegan, John, and Andrew Wheatcroft. *Who's Who in Military History*. New York: William Morrow, 1976, p. 95.

3. Ibid., p. 322.

4. Young, Peter. *Great Battles of the World*. New York: Bookthrift, p. 122.

5. Messenger, *Blitzkrieg Story*, p. 38.

6. Ibid.

7. Ibid., p. 42.

8. Keegan, *Second World War*, p. 421.

9. Ibid.

10. Taylor, A. J. P. *The Second World War*. New York: Putnam's, 1975, p. 66.

11. Keegan, *Second World War*, p. 91.

12. Ibid., p. 92.

13. Ibid., p. 432; Taylor, *Second World War*, p. 78, says 30,000, but he also says "during the Blitz" and "mostly in London." A lot of people, in any case.

14. Keegan, *Second World War*, p. 432.

15. Ibid., p. 420.

16. Taylor, *Second World War*, p. 129.

17. Keegan, *Second World War*, p. 430.

18. Ibid., pp. 432–33.

19. Taylor, *Second World War*, p. 179.

20. Sulzberger, C. L. *The American Heritage Picture History of World War II*. New York: American Heritage, 1966, p. 589.

21. Ibid., p. 608.

22. Taylor, *Second World War*, p. 226.

20 High Tech

1. Murray, Williamson and Robert H. Scales, Jr. *The Iraq War: A Military History*. Cambridge, MA: Belnap, 2003, pp. 1–4.

2. For information on most of the new weapons used in sea, land, and air warfare, see Diagram Group, *Weapons*.

3. Murray and Scales, *Iraq War*, p. 264.

4. Ibid., p. 265.

21 Low Tech

1. Lincoln, W. Bruce. *Red Victory: A History of the Russian Civil War*. New York: Simon and Schuster, 1989, p. 32.

2. German combat deaths of 4 million amounted to 5 percent of the country's entire population. Soviet combat deaths of 7 million were 3.8 percent of the population.

3. Currey, Cecil B. *Edward Lansdale: The Unquiet American*. Boston: Houghton Mifflin, 1988, pp. 146–47.

4. Van Crevald, *Technology and War*, p. 181.

5. Mao Tse-tung. *Mao Tse-tung on Guerrilla Warfare*. New York: Praeger, 1961.

6. Guevara, Che. *Guerrilla Warfare*. New York: Random House, 1961, pp. 8–12.

7. For a quick summation of this unusual war, see Weir, *Soldiers in the Shadows*, pp. 143–61.

8. Murray and Scales, *Iraq War*, p. 255.

Index

Abatis, defined, 265
Abrams tanks, 251
Adams, Henry, 161
Admiral, defined, 265
Advance by rushes, 170
Aerial observation, 126, 137–38
Afghanistan, 253–54
Afrikaner revolt, 149–55
Agar machine guns, 138
Agriculture, feudalism and, 36–41
Air forces (air power), 221–27,
    247–49, 253. See also Royal Air
    Force
    realization of, 224–27
Akagi, 231–32
Albertus Magnus, 56
Aleutian Islands, 228–29
Alexander III (the Great), 18, 19, 24,
    28, 75
Allied shipping, 188, 190–91,
    192–200, 202–5
Alligator, 192
Almeida, Francisco de, 78–79
American Civil War, 125–40
    ironclads and, 127, 128–33, 158
    machine guns and, 138–40
    rifles and, vii–viii, 127–28,
        133–38, 188
    Sheridan and, 180–81
    timeline, xvi
    weapons drill, 92
American colonies, 85–86, 111–12
American Revolution
    columns and, 122
    Ferguson and, 114–15
    France and, 118–19
    rifles and, 109, 134
    soldiers and, 112
Anstruther, P. R., 149–55
Archerfish, 205, 232
Argonaut, 190
Arizona, 222–23

Armies, 117–24. *See also specific*
    armies
    of automatons, 107–12
    defined, 265
    private, xi–xii, 43–49
    standing, xiii, 83–104
Armor, 8–9, 13–17, 26, 49, 251–52
Armored personnel carriers (APCs),
    252
Army groups, defined, 265
Arnim, Hans Georg von, 96,
    99–100, 102–3
Arrows, 2–4, 8–9, 13–14, 25, 28–29,
    31, 49
Artillery, 106–7, 113, 115
    Iraq War and, 250–52
    Macedonians and, 18–19
    World War I and, 170–76
Aryans, 30
Asdic, 197–98, 202–3
Atlantic, Battle of, 196–99, 203
Atlantic Fleet, 226–27
Atlatls, 7–8
Attila the Hun, 31
Austria-Hungary, 141–47
Austrian Succession, War of, 105–7,
    111
Automatics, 171, 174, 177, 265
Ávalos, Fernando Francisco de, 65
Axes, 2–3, 7–8
Aztecs, 7–8, 9

Bacon, Roger, 56–57
Balance of terror, 176
Balloon corps, 126, 137–38, 188
Bamboo guns, 53
Barbarians, 34–41, 75
Barracks, 108
Barrett, John, 164
Bar shots, 128–29
Barton, Clara, 127
Battalions, 91, 265